Green Energy and Technology

Climate change, environmental impact and the limited natural resources urge scientific research and novel technical solutions. The monograph series Green Energy and Technology serves as a publishing platform for scientific and technological approaches to "green"—i.e. environmentally friendly and sustainable—technologies. While a focus lies on energy and power supply, it also covers "green" solutions in industrial engineering and engineering design. Green Energy and Technology addresses researchers, advanced students, technical consultants as well as decision makers in industries and politics. Hence, the level of presentation spans from instructional to highly technical.

Indexed in Scopus.

Indexed in Ei Compendex.

Vahid Vahidinasab • Behnam Mohammadi-Ivatloo
Editors

Electric Vehicle Integration via Smart Charging

Technology, Standards, Implementation, and Applications

Springer

Editors
Vahid Vahidinasab
Department of Engineering, School of Science and Technology
Nottingham Trent University
Nottingham, UK

Behnam Mohammadi-Ivatloo
Electrical and Electronics Engineering Department
Muğla Sıtkı Koçman University
Muğla, Turkey

ISSN 1865-3529 ISSN 1865-3537 (electronic)
Green Energy and Technology
ISBN 978-3-031-05908-7 ISBN 978-3-031-05909-4 (eBook)
https://doi.org/10.1007/978-3-031-05909-4

This Springer imprint is published by the registered company Springer Nature Switzerland AG
The registered company address is: Gewerbestrasse 11, 6330 Cham, Switzerland

Preface

Electric vehicles (EVs) are looked upon as a potential solution to cope with the environmental effects of the transport sector, which is an energy-intensive sector worldwide with a significant share in the emission of pollutants. EVs can offer new opportunities for both the public sector and also consumers if they are integrated and managed in a smart way. Electric vehicles smart charging (EVSC) is an important requirement of the current evolving situation, and it would enable consumers to benefit from cheaper electricity during off-peak periods, that is, when electricity demand is low, and in parallel can avoid causing unnecessary reinforcement of the electricity network. Also, by EVSC, it is possible to manage them to be shifted to the periods where there is abundant renewable electricity generation in the network for helping to increase the rate of decarbonization.

The book covers a comprehensive study on electric vehicles' smart charging and its role in the decarbonization efforts around the world for meeting long-term strategies. It is expected to present practical aspects and approaches with evidence from applications to real-world energy systems. By providing in-depth technical discussions, use cases, and examples, understanding of the presented methods is made easier.

This book is an attempt to bring together experts from the different disciplines related to the EVSC topic as we strongly believe that there is a wealth of knowledge available in each discipline that is not widely known in the other disciplines and could be usefully employed to overcome the challenges we are facing now and provide a comprehensive and in-depth reference.

Chapter 1, entitled "Standardized Domestic EV Smart Charging for Interoperable Demand Side Response: PAS 1878 and 1879," mentions that, recently published by the British Standards Institution, PAS 1878 and PAS 1879 present a standardized technical framework for energy-smart appliances, including smart electric vehicle charging points, to provide demand-side response services to domestic consumers and grid-side actors. The framework covers electric vehicle smart charging based on time of use tariffs, as well as aggregated flexibility services. The framework was developed to deliver the four principles of interoperability, data privacy, cyber security, and grid stability.

Chapter 2, entitled "Concept of the Control Strategies for Li-Ion Batteries to Improve Reliability in the Electric Vehicle (EV) Applications," studies the impact of the nonlinear features of lithium-ion batteries on the lifetime performance, reliability assessment, and control of them. In this chapter, the BMS definition, SoH and SoC methods, and battery fault detection methods have been described as key aspects of the control strategy of Li-ion batteries for improving the reliability of the system. Moreover, the challenges and further work relating to the estimation of the state of function of the Li-ion batteries for EV applications have been investigated.

Chapter 3, entitled "Recognition of Electric Vehicles Charging Patterns with Machine Learning Techniques," aims to investigate the charging behavior of EVs. By analyzing the charging patterns of the EV stations, different rules could be developed to manage the EV charging patterns. This chapter introduced the machine learning (ML)–based approach to cluster the EVs charging behaviors. Identifying clusters of EVs charging patterns is conducted via the unsupervised learning ML method, while the supervised learning ML method was utilized for further classification of the dataset.

Chapter 4, entitled "Cybersecurity and Data Privacy Issues of Electric Vehicles Smart Charging in Smart Microgrids," introduces that penetration of EVs in smart microgrids makes it essential to manage their charging patterns and provide some ancillary services such as peak-load reduction, congestion management, frequency regulation, management of uncertainties, and zero exhaust emissions. These functionalities depend on the cyber-physical security of data collected from various EVs management centers. This chapter addresses the cyber-physical challenges of the EVs smart charging systems, the attack patterns and impacts in power grids, and the attacker-defender model.

Chapter 5, entitled "Evaluation of Cyberattacks in Distribution Network with Electric Vehicle Charging Infrastructure," studies the cyberattack risk and security in EV systems. The security and resilience of smart grids are under the risk of cyberattacks due to their well-equipped communication and telecommunication systems and technologies, as well as their interconnected parts and infrastructures. The electric vehicle infrastructure (EVI) infrastructure is an ideal target for such attacks for many reasons, including the mobile nature of electric vehicles. The electric vehicles, as the moving part of smart grids, become compromised due to a cyberattack, and it can be propagated by other components of the infrastructure, such as electric vehicle supply equipment (EVSE).

Chapter 6, entitled "Electric Vehicle Services to Support the Power Grid," discusses the services provided by the optimal charging and discharging of EVs to the power grid. The mathematical modelling of the charging and discharging optimization problem of EVs is also considered.

Chapter 7, entitled "Smart Charging of EVs to Harvest Flexibility for PVs," investigates a smart charging model of EVs in order to facilitate efficient integration of EVs and PVs in distribution networks. A novel flexibility factor is defined for EVs demand profile that can facilitate a range of modified charging profiles to realize EVs' smart charging behavior. The proposed model and obtained results imply that for the modified EVs' behavior, the existing network's capability to supply more EVs and PVs can increase significantly.

Chapter 8, entitled "A Robust Optimization-Based Model for Smart Charging of PEV Under Multiple Uncertainties," proposes a robust PEV charging model wherein the optimal charging/discharging strategy of the PEV is determined considering the worst-case possible deviation of the uncertain variables from their anticipated values throughout their pertinent robust regions.

Chapter 9, entitled "The Role of Smart Electric Vehicle Charging in Optimal Decision-Making of the Active Distribution Network," evaluates the impact of electric vehicles high penetration on the active distribution network. In this chapter, the impact of electric vehicles smart charging strategy and dumb charging on the optimal decision of the active distribution network has been investigated. A stochastic-robust hybrid approach is proposed to evaluate the effect of uncertainty parameters on the optimal decision of the active distribution network.

Chapter 10, entitled "Operational Challenges of Electric Vehicle Smart Charging," is discusses the potentials of smart charging in the operation of electric energy systems, status quo, and its future challenges.

The book is a multidisciplinary reference that is suitable for both dedicated and general audiences that include researchers and industry stakeholders who have focused on the field of electric vehicles smart charging as well as researchers and developers from different branches of engineering, energy, transportation, economics, and operation research fields. It is expected that readers be graduates of science and engineering degree programs having a basic mathematical background.

The material provided in this book fits the needs of advanced undergraduate or graduate modules and courses on energy systems. The book offers an adequate mixture of technology and engineering background and modelling approaches that makes it a suitable reference for the students as well as researchers and engineers in academia and industry who are active in the field.

To conclude, we would like to sincerely thank all of the authors who contributed to this book and also all the reviewers for their insightful observations and helpful comments.

Also, the editors would like to extend their deep gratitude to the following reviewers (sorted alphabetically) for their thoughtful comments on all the submitted book chapters including those that were accepted and published in this book:

Hamdi Abdi
Masoud Ahmadigorji
Sajad Najafi-Ravadanegh
Omid Sadeghian
Farkhondeh Jabari
Nina Klein
Amin Mansour-Saatloo
Majid Mehrasa
Arash Moradzadeh
Younes Noorollahi

Arman Oshnoei
Zoya Pourmirza
Abbas Rabiee
Mehdi Zeraati

The editors, authors, and reviewers of this book have dedicated their time and enthusiasm to creating it in the hope that it will be useful to researchers, graduate students, and practitioners interested in this field.

Nottingham, UK Vahid Vahidinasab
Muğla, Turkey Behnam Mohammadi-Ivatloo

Contents

Editors and Contributors

About the Editors

Vahid Vahidinasab, PhD is an Senior Electrical Power Engineering Lecturer and Director of GridLab at Nottingham Trent University, Nottingham, UK where he leads teaching and research in the area of sustainable power and energy systems. Before that, he was an Assistant Professor at Shahid Beheshti University, Tehran, Iran, and a Senior Research Associate in the EPSRC National Centre for Energy Systems Integration at Newcastle University, Newcastle upon Tyne, UK, where he worked with the EU Horizon 2020 inteGRIDy project and Active Building Centre (ABC) Research Programme. He has demonstrated a consistent track record of attracting external funds and managed industrial projects and closely worked with 12 large and complex national/international projects. From 2011 to 2018, he held a number of leadership roles at SBU and the Niroo Research Institute. His research works have been funded by the UK EPSRC, EU-H2020, Horizon Europe, industry partners and local utilities. His research interests include power and energy systems modelling, analysis, operation and control, energy systems integration, power systems and grids resilience, smart grids and microgrids, and distributed energy systems. Dr Vahidinasab is the Editor-in-Chief of Springer Smart Grid and Sustainable Energy, an Associate Editor for the IEEE Transactions on Industry Applications. He is also a member of the Editorial Board and a Subject Editor of the IET Generation, Transmission and Distribution, and an Associate Editor of the IET Smart Grid, Nature Scientific Reports, and IEEE Access.

Behnam Mohammadi-Ivatloo, PhD is a member of the Faculty of Engineering in the Department of Electrical and Electronics Engineering at Muğla Sıtkı Koçman University, Turkey. He was previously a senior research fellow at Aalborg University, Denmark. He is also a professor at the University of Tabriz, from where he is currently on leave. Before joining the University of Tabriz, he was a research associate at the Institute for Sustainable Energy, Environment and Economy at the University of Calgary. He obtained MSc and PhD degrees in electrical engineering from Sharif University of Technology. His main research interests are renewable energies, microgrid systems, and smart grids.

Contributors

Hamdi Abdi Electrical Engineering Department, Engineering Faulty, Razi University, Kermanshah, Iran

Saleh Aghajan-Eshkevari Department of Electrical Engineering, Shahid Beheshti University, Tehran, Iran

Raziye Aghapour Department of Electrical Engineering, Abbaspour School of Engineering, Shahid Beheshti University, Tehran, Iran

Davood Ahmadian Faculty of Mathematics, Statistics and Computer Sciences, University of Tabriz, Tabriz, Iran

Masoud Ahmadigorji Department of Electrical Engineering, Nour Branch, Islamic Azad University, Nour, Iran

Mohammad Taghi Ameli Department of Electrical Engineering, Shahid Beheshti University, Tehran, Iran

Hamidreza Arasteh Power Systems Operation and Planning Research Department, Niroo Research Institute (NRI), Tehran, Iran

Sasan Azad Department of Electrical Engineering, Shahid Beheshti University, Tehran, Iran

Seddik Bacha University of Grenoble Alpes, CNRS, Grenoble INP, G2ELAB, Grenoble, France

Uldouz Bazazi Resilient Smart Grids Research Lab, Electrical Engineering Department, Azarbaijan Shahid Madani University, Tabriz, Iran

Foad H. Gandoman Faculty of Electrical and Computer Engineering, University of Tabriz, Tabriz, Iran

Lee Gould Kinberry Limited, Glastonbury, UK

Ahmad Hably University of Grenoble Alpes, CNRS, Grenoble INP, GIPSA-Lab, Grenoble, France

Farkhondeh Jabari Power Systems Operation and Planning Research Department, Niroo Research Institute (NRI), Tehran, Iran

Andrew Keane School of Electrical and Electronic Engineering, University College Dublin, Dublin, Ireland

Nina Klein Department of Business, Energy and Industrial Strategy, Westminster, London, UK

Ian Llewellyn Department of Business, Energy and Industrial Strategy, Westminster, London, UK

Mousa Marzband Northumbria University, Electrical Power and Control Systems Research Group, Newcastle upon Tyne, UK

Majid Mehrasa University of Grenoble Alpes, CNRS, Grenoble INP, G2ELAB, Grenoble, France

Behnam Mohammadi-Ivatloo Electrical and Electronics Engineering Department, Muğla Sıtkı Koçman University, Muğla, Turkey

Mansour Moradi Department of Electrical Engineering, Technical and Vocational University (TVU), Kermanshah, Iran

Nima Nasiri Resilient Smart Grid Research Lab, Electrical Engineering Department, Azarbaijan Shahid Madani University, Tabriz, Iran

Vahid Nasiriyan Faculty of Electrical and Computer Engineering, University of Tabriz, Tabriz, Iran

Abbas Rabiee School of Electrical and Electronic Engineering, University College Dublin, Dublin, Ireland

Sajad Najafi Ravadanegh Resilient Smart Grid Research Lab, Electrical Engineering Department, Azarbaijan Shahid Madani University, Tabriz, Iran

Maryam Shahbazitabar Electrical Engineering Department, Engineering Faulty, Razi University, Kermanshah, Iran

Mohammadreza Sheibani Power Systems Operation and Planning Research Department, Niroo Research Institute (NRI), Tehran, Iran

Mohammadreza Shekari Power Systems Group, Department of Electrical & Computer Engineering, Tarbiat Modares University, Tehran, Iran

Alireza Soroudi School of Electrical and Electronic Engineering, University College Dublin, Dublin, Ireland

Vahid Vahidinasab Department of Engineering, School of Science and Technology, Nottingham Trent University, Nottingham, UK

Mehdi Zeraati Power Systems Operation and Planning Research Department, Niroo Research Institute (NRI), Tehran, Iran

Saeed Zeynali Resilient Smart Grid Research Lab, Electrical Engineering Department, Azarbaijan Shahid Madani University, Tabriz, Iran

Chapter 1
Standardised Domestic EV Smart Charging for Interoperable Demand Side Response: PAS 1878 and 1879

Lee Gould, Nina Klein, and Ian Llewellyn

1.1 Introduction

1.1.1 Purpose of Demand-Side Response

The transition to net zero greenhouse gas emissions presents two substantial changes for the electricity system. Firstly, electricity supply will be more variable, coming predominantly from renewable energy sources such as wind and solar. Secondly, electricity demand will increase, with the electrification of domestic heating and transport.

These changes require a fundamental shift in how the electricity system is operated. Previously, fossil fuel generation was centralised and dispatchable, so supply could be increased or decreased minute by minute to match the demand on the system. In the future, however, renewable generation will be predominantly non-dispatchable, generating, for example, when the sun shines or wind blows, so instead, demand must be made flexible to match the variable supply on the system whilst operating within the constraints of the physical network.

The net zero transition therefore requires new flexible technologies on the demand side of the electricity system that were not required with fossil fuel generation. This flexibility could be provided in several ways:

L. Gould
Kinberry Limited, Glastonbury, UK
e-mail: lee.gould@kinberry.com

N. Klein (✉) · I. Llewellyn
Department of Business, Energy and Industrial Strategy, Westminster, London, UK
e-mail: nina.klein@nationalgrideso.com; ian.llewellyn@beis.gov.uk

V. Vahidinasab, B. Mohammadi-Ivatloo (eds.), *Electric Vehicle Integration via Smart Charging*, Green Energy and Technology,
https://doi.org/10.1007/978-3-031-05909-4_1

- Allowing system operators to control the operation of some types of appliances within bounds set by consumer wishes, known as "demand-side response" (DSR) services
- Interconnecting many networks together in a grid to produce bigger networks, where extra demand in one part of the super-network is likely to be met by surplus power in another
- Electrical storage systems, whether lithium-ion battery systems, hydroelectric systems or new forms of storage being developed such as power to gas

In the UK, analysis suggests that increased flexibility provided through DSR, electricity storage and interconnection could reduce the decarbonised electricity system costs by £6–12 billion per year by 2050 [1].
DSR is a form of flexibility which can be used for many purposes:

- To match electricity demand to the amount of supply available on the market (e.g. through time of use tariffs) so allowing consumers to reduce their energy bill
- To allow electricity system operators to adjust short-term imbalances between supply and demand on a second-by-second basis to allow grid balancing without dispatchable fossil fuel generation
- To allow consumers to choose to decrease their carbon footprint by automatically running their appliances at times of lowest grid carbon intensities (e.g. through instantaneous grid carbon intensity data)
- To manage peak loads on the electricity distribution and transmission networks, to reduce the need for future network upgrades to cope with the electrification of heat and transport
- To control distribution or transmission network characteristics, such as frequency, voltage and inertia, to support system operation
- To provide an emergency response function in event of a fault on a generating asset or network, to shed load quickly to prevent more widespread blackouts

In this chapter, we describe a standardised framework for providing secure and interoperable DSR services. The framework was developed in the UK with the British Standards Institution as two technical standards: PAS 1878 and PAS 1879. The specific use case described in this chapter is for controlling domestic electric vehicle (EV) charging. Smart EV charging is ideal for providing DSR services, as the consumer has specific requirements (to be able to drive the EV at a specific time) but is less concerned about precisely when the EV is charged provided their requirements are met.

1.1.2 Status Quo, Challenges and Outlook

Several studies have looked at the effect of EVs on the electricity network with and without smart charging, both through computer modelling studies and small-scale trials. A global review of smart charging practices was recently published [2].

Since it is generally assumed that EV users will have the ability to override the smart functionality at any time (so can opt-out of smart charging), other studies have looked at the incentivisation of consumers needed to move appreciable load from peak to off-peak times. For example, a recently published study, using over 400 Northern California households in the largest pilot project of its type, showed that 15–20% of charge can be shifted out of any given hour and 20–30% of charge can be shifted into any given hour by relatively modest levels of incentivisation coupled with automated smart technology [3].

The landscape for EV smart charging standards and protocols is currently diverse, with various challenges to be overcome. Whilst some open standards for smart charging exist [4], they are not as yet formally agreed standards at international standards bodies and do not have an openly accessible governance process. There are some emerging international standards, e.g. IEC 63110-1 and IEC 63110-2 and ISO 15118-20 (see Sect. 1.4.1); however, they are not complete, and some are likely to require years of further development. In this absence of commonly agreed international standards across EVs and Chargepoints, industry players have developed their own proprietary bespoke solutions. This is problematic as proprietary approaches do not deliver interoperability and its associated benefits.

The various current EV smart charging standards and protocols often have drawbacks in terms of their approach or the functionality they provide. Many extant approaches are specific to EVs only, despite the fact that DSR can be provided by a range of appliances with flexible energy consumption. PAS 1878 and PAS 1879 provide a standardised technical framework applicable to all energy smart appliances (ESAs), including smart EV Chargepoints, with any specific requirements for an individual appliance type specified in PAS 1878 Clause 8. Additionally, whilst existing standards and protocols can deliver "smart" functionality, they often neglect other key operational aspects important to system actors such as interoperability, data privacy, cyber security and grid stability. The framework in the PAS standards was developed to deliver all four of these key principles for smart charging. The PAS standards additionally facilitate sophisticated smart charging functionality, using forecast "power profiles" (see Sect. 1.3.2), not just simple on/off commands, and focus on delivering real-time requests for DSR services not just optimising schedules in advance.

Given the mass deployment of EVs which is forecast, there is a need to ensure that charging is smart to minimise the economic impacts to upgrading generation and network infrastructure. It is important that these EVs and Chargepoints operate according to agreed international standards, both to deliver consumer benefits and to support a coordinated and rapid transition to net zero carbon. The PAS standards provide an architecture which gives interoperability whilst allowing appliance-specific protocols where necessary, and they provide a framework for operation which enables sophisticated energy smart operation across different appliance types.

Various innovative trials of EV smart charging have been undertaken in recent years [5–9]. At best these trials use open standards and at worst they use proprietary protocols, but in both cases, the approaches are specific to EVs only and they often don't facilitate the sophisticated DSR functionality required by grid-side actors in the electricity system. It is hoped that future trials will implement PAS 1878 and PAS 1879 in smart EV Chargepoints and back-end DSR operation platforms. This will be a valuable next step to test the PAS standards in real-world settings, to see the benefits the standards can deliver and suggest any feedback for future updates.

1.1.3 Assumptions of the Standardised Framework

The electricity system contains different types of actors, who depending on their legal status are allowed (or prohibited) from instigating different types of DSR. The standardised framework considers two key actor types:

- *System Operators*

  These are regulated entities, such as transmission system operators and distribution system operators. Legally they cannot participate in the wholesale electricity market, cannot supply electricity directly to consumers (or directly set consumer prices) and are not allowed to own generation assets. However, they can operate ancillary markets for grid balancing and operation purposes; through this mechanism, they contract the provision of specific DSR services.
- *Market Participants*

  These are licenced entities, such as suppliers and generators, which are more lightly regulated than system operators. Market participants can participate in the wholesale electricity market and can supply electricity directly to consumers, including directly setting consumer prices. Hence, through time of use tariffs (TOUT), they can incentivise the provision of DSR services.

The standardised framework considers that an EV Chargepoint can provide DSR services if it has "energy smart appliance" (ESA) functionality, as specified in PAS 1878 [10]. To provide DSR services, the Chargepoint must have the ability to:

- Receive external signals (incentives or requests) from data sources or actors in the electricity system.
- Modulate (increase/decrease or shift in time) the electricity consumption/production of the Chargepoint in response to those external signals.

All DSR services are undertaken in line with consumer preferences, which the Chargepoint includes when creating flexibility options. The Chargepoint must therefore have an interface, either in the equipment or remotely, e.g. via a mobile phone app, in order to receive consumer preferences.

Note: Throughout this chapter, the term "Chargepoint" is used to mean an ESA which is an EV chargepoint. The term "Chargepoint" is used when a specific example is being given, and the term "ESA" is used when a general principle or concept is being presented.

There are two types of DSR services, set out in the standardised framework, which EV smart charging can provide:

- *Routine Mode DSR*
  The Chargepoint operates based on incentives which are generally set in advance:
  - The incentives may be set by electricity system market participants, based on multiparty market signals. An electricity supplier bids in wholesale markets and then sets a TOUT 24 hours in advance as an incentive.
  - As an alternative, the Chargepoint can download forecast grid carbon intensity data from the Internet or obtain local PV generation forecasts and use these incentives to optimise its operation to lower carbon emissions.
- *Response Mode DSR*
  The Chargepoint operates based on requests which are generally made in near real time:
  - The requests may be initiated by electricity system market participants or system operators, based on firm bilateral contracts for the delivery of DSR services such as frequency response, balancing or constraint management.
  - An intermediary called a "DSR service provider" (DSRSP) is contracted to deliver these DSR services. The DSRSP also (directly or indirectly) has a contract with consumers to allow control of individual ESAs, subject to always meeting consumer wishes.
  - The DSRSP, therefore, sends near real-time requests to an aggregated group of Chargepoints to deliver the DSR services.

PAS 1878 and 1879 provide a framework encompassing both routine and response mode, whilst specifying response mode in detail [10, 11]. A hierarchy of modes is described, with response mode taking priority over routine mode. This is sensible because failure to achieve minute-by-minute control of demand/supply or frequency in a decarbonised network using response mode services will lead to circuit breakers tripping and widespread disconnection, whereas controlling demand to match the market availability of supply simply allows more efficient economic operation of the electricity network but does not lead to disconnection. Note that routine mode TOUT incentives may only be set by market participants and must operate on 30-minute or 1-hour time periods. Response mode is a more universal way of delivering DSR services; there are no legal restrictions as to who may instigate it and no timescale restrictions for the same reason.

The standardised framework also includes cyber security requirements. This is because, in aggregate, large numbers of ESAs connected to the electricity network present a risk to critical infrastructure. Therefore, the cyber security requirements in the framework go beyond basic Internet of things security approaches, whilst being proportionate to risks and respecting compromises between cost, usability and security. In addition to the specific technical requirements of the framework, national authorities may choose to implement additional measures, e.g. approval for DSRSPs to operate; see Sect. 1.2.2.9.

1.1.4 Overview of Operation

The high-level architecture is shown in Fig. 1.1, which depicts the logical functionalities in the DSR framework for EV smart charging. The need for DSR services comes from the system operators and market participants (SOMPs), e.g. transmission or distribution system operators and balance responsible parties, who require DSR for the various purposes outlined in Sect. 1.1.1. SOMPs contract DSR service providers (DSRSPs) to deliver these DSR services from domestic appliances, including smart EV chargepoints. The DSRSP connects to appliances from different manufacturers interoperably via a Customer Energy Manager (CEM), a logical functionality (which can be cloud or edge based) that translates flexibility information between the DSRSP and the ESA. (Note: The CEM may optionally connect to a Home Energy Management System (HEMS), which can provide additional optimisation DSR services across multiple appliances in a premises.) The CEM communicates this flexibility information with the ESA, a physical device which operates flexibly to deliver DSR services. For EV smart charging, the ESA functionality is delivered by the EV Chargepoint which connects to the EV (the vehicle). The user is also a key functionality, as user preferences are input to the CEM or Chargepoint to form the basis of flexibility offers.
The architecture has been designed with a number of key features in order to deliver user, service provider and system benefits:

- At a minimum, an ESA (the Chargepoint) must be supplied with a CEM, but it can subsequently swap to another CEM (or HEMS). This ensures the ESA is interoperable with all DSRSPs when delivered by the manufacturer, whilst also allowing third parties to innovate with consumers to customise their ESA configuration if desired.
- The modular architecture of individual interoperable ESAs allows multiple DSRSPs per premises. Firstly, this facilitates DSRSP specialisation; rather than being forced to operate all ESAs in a premises, one DSRSP can specialise in Chargepoints, whilst another DSRSP can operate, for example, heat pumps. Secondly, this benefits the consumer who can "shop around" and subscribe each ESA individually to whichever DSRSP offers the best price or service for their flexibility.
- The energy smart functionality or "intelligence" is placed within the ESA (not with a different service provider); so the Chargepoint will determine its own operation schedule and flexibility options. This is pragmatic as only the appliance itself and manufacturer know its operational details; it also ensures the appliance won't go "dumb" on change of service provider. There are additional resilience

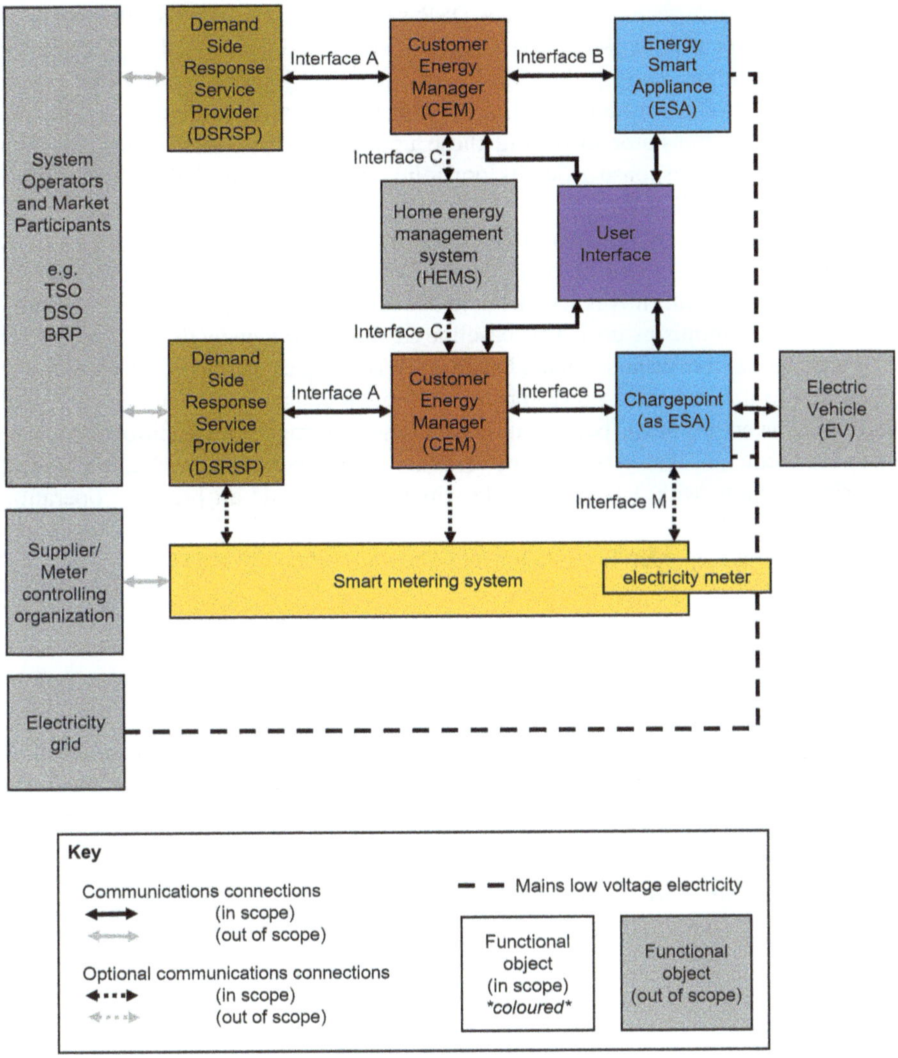

Fig. 1.1 Architecture of EV smart charging logical functionalities for DSR, showing the overall structure

benefits when the ESA can operate without a cloud connection, and there are consumer data privacy benefits when their data does not leave the edge device.

The system operation is based on the Chargepoint intelligently creating flexibility options for the DSRSP in the form of multiple "power profiles" which are operational forecasts of power vs time that deliver user preferences. A push-update model is utilised, where the Chargepoint creates (or updates) power profiles and sends them to the CEM, which translates them and offers them to the DSRSP over a standardised interface. To deliver a DSR service, the DSRSP can then request a specific power profile, which the CEM translates and sends to the Chargepoint for implementation. During this DSR event, the Chargepoint monitors its power consumption/production and updates its power profiles as necessary. Again, the CEM acts as a translator providing these updates to the DSRSP who can then call more or less response from its portfolio of ESAs to fulfil the DSR service requirement of the SOMP.

The system also contains an operational hierarchy, based on the types of DSR described in Sect. 1.1.3. As a baseline the Chargepoint operates in routine mode, optimising against incentives such as time of use tariffs in order to achieve consumer goals such as minimising cost. This baseline can be overridden by the higher-priority response mode; this occurs when a DSRSP sends a request and the ESA changes its operation to the requested power profile. Although consumer preferences are built into the power profile in both routine and response mode, an additional manual override mode takes priority over any response mode operation. Finally, the highest priority is safety with a failsafe mode which can override all previous operation modes.

The step-by-step operation sequence is shown at a high level in Fig. 1.2.

1.1.5 Underpinning Principles

The technical architecture presented aligns with the four principles outlined below. The principles are important to enable the benefits of smart charging, both for consumers and the electricity system, whilst also protecting against possible risks.

Interoperability

The ability of an ESA to work seamlessly across any appropriate DSR service operated by any authorised party:

- The overall architecture facilitates interoperable interfaces and specifically, DSRSP-ESA interoperability is specified. This allows consumers to switch an ESA to a different DSRSP, preventing lock-in and enabling a competitive market for DSR services.
- DSRSP-ESA interoperability is enabled through the specification of a data model and communications protocol (which is based on an open standard), along with a defined common operating framework with minimum functionality and security requirements. Therefore, the DSRSPs only need to offer one standard interface to all smart devices in order to enrol them in DSR services.

Data Privacy

The secure transmission and storing of data on the device or with any controlling party:

- The overall operating framework minimises data transfer, and consumers are in control of data at all times, with clear consent procedures.
- Only the data necessary to operate the DSR service is transferred to the DSRSP, communications are authenticated and encrypted and storage is secured with secure removal procedures.

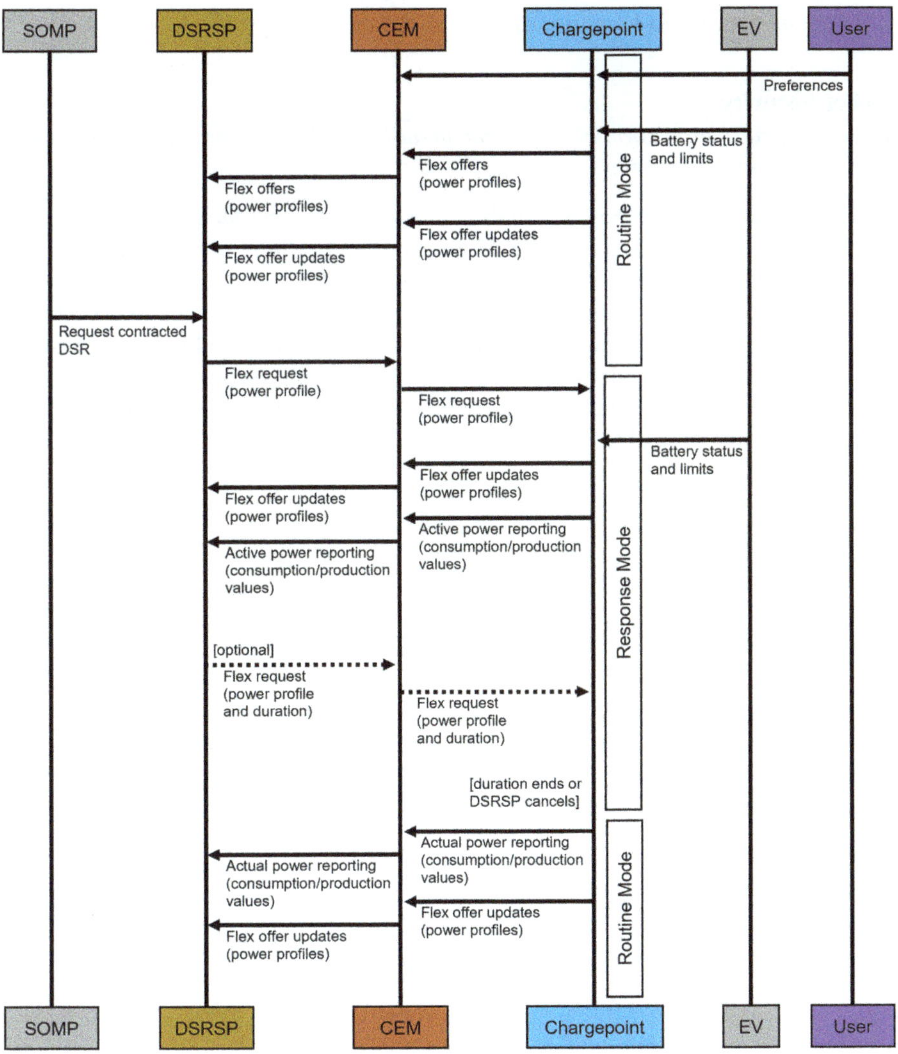

Fig. 1.2 Operational sequence of EV smart charging for DSR

Grid Stability

The prevention of outages on the grid caused by inappropriate operation of ESAs:

- The operating framework and technical requirements are specified so that ESAs would not inadvertently represent a risk to grid stability.
- ESAs are required to provide up-to-date flexibility information and power consumption reporting, to enable DSRSPs to act on accurate information. An operational hierarchy is defined to deliver SOMP needs and consumer wishes. Randomised offsets must be applied to certain operations to avoid steep ramp rates, and a process is defined for exception conditions, including loss of communications, to minimise disruption to grid stability.

Cyber Security

The appropriate protection of ESAs from unauthorised access and the correct use of ESAs by authorised parties:

- The operating framework specifies a proportionate approach to cyber security, covering both the ESA and DSRSP, the communications sent and received and the communication networks used.
- Minimum security requirements are set for information storage and exchange (including encryption), authentication of parties (based on public key infrastructure (PKI)), secure device configuration and operation, secure software/firmware updates and secure data management. Additionally, existing cyber security standards are applicable.

1.1.6 Scope

The technical framework presented provides a common platform to enable various interoperable DSR services, whilst focusing on response mode services in particular. The framework deliberately provides only a *minimum* specification, in order to meet the four principles outlined in Sect. 1.1.5 whilst leaving room for innovators to build additional services on top.
The following items are included in the technical framework:

- The DSR functional requirements of an ESA (the Chargepoint)
- An overall DSR operation framework
- A DSR system architecture with defined objects and interfaces
- A data model and communications protocol for response mode DSR

The following items are excluded from the technical framework:

- Appliance functionality not related to DSR
- General appliance safety

- General consumer protection
- Contracting or payment aspects

The technical framework is compatible with the following:

- The EU Smart Grid Architecture Model (SGAM) framework
- The system architecture and functionality described in the outputs of the following technical committees:
  - CENELEC TC205 "Home and Building Electronic Systems"
  - CEN/CENELEC TC59X "Performance of household and similar electrical appliances"
  - IEC TC57 "Power systems management and associated information exchange"
  - IEC TC69 "Electrical power/energy transfer systems for electrically propelled road vehicles and industrial trucks" (specifically JWG1 for 15118 and JWG11 with TC57 for 63110)
- Home Energy Management Systems (HEMS), through interface C which could be defined for interoperability
- Smart metering systems, through interface M which could be defined for interoperability (for GB smart metering, this is defined in PAS 1878 Annex D)

The technical framework encompasses the following types of EV smart charging, at different levels of detail:

- *Optimisation for local PV or battery integration and optimisation against a TOUT or grid CO2 intensity*. These forms of smart charging fall under routine mode DSR. This is fully integrated in the technical framework, but is not fully specified (due to asset- or country-specific data inputs and there being no "optimum" optimisation algorithm). However, PAS 1878 Annex D provides a specification for the provision of TOUTs via the GB smart metering system [10].
- *Aggregated control for transmission or distribution network services or balancing market services*. These forms of smart charging fall under response mode DSR. This is fully integrated in the technical framework and fully specified for interoperability between DSRSPs and Chargepoint.
- *Scheduling for consumer convenience*. This form of smart charging falls under routine mode DSR, but in its most basic form (based on user preferences only and not considering any energy system incentives). Again, this is fully integrated in the technical framework, but it is not specified how this must be done (rather this is left to individual Chargepoint manufacturers); it is simply specified that user preferences must be collected and met in all modes of DSR operation.

Whilst this chapter focuses on a Chargepoint as the ESA, other domestic appliances can also fulfil the ESA functional requirements. PAS 1878 covers:

- Heating, ventilation and air conditioning (HVAC) appliances
- Cold appliances, e.g. fridge freezers
- Wet appliances, e.g. washing machines, dishwashers

- Battery storage
- EV Chargepoints

The ability of PAS 1878 to cover other appliance types, as well as Chargepoints, minimises the costs of enrolment for DSRSPs and so allows consumers to achieve higher revenues for contracting their flexibility management to DSRSPs.

1.2 System Architecture

1.2.1 Functional Architecture

1.2.1.1 Compatibility with International Standards

The PAS functional architecture is built to be compatible with international smart grid and EV standards, including the SGAM.

The architectural foundation is based on the smart grid architecture used by CENELEC TC205, CENELEC TC59X and IEC TC57, which facilitates the energy management of devices in the home and connection with grid-side actors. The semantic mapping of functional components is shown in Fig. 1.3.

PAS 1878	**CENELEC/IEC**
Energy Smart Appliance (ESA)	Smart Appliance
Customer Energy Manager (CEM)	Customer Energy Manager Resource Manager or Customer Connectivity and Energy Manager
On-premises router/gateway, mobile data modem on ESA or CEM etc.	Energy Management Gateway
DSR Service Provider (DSRSP)	Actor A (smart grid)
Smart metering system Supplier/Meter controlling organisation	Smart metering system Actor B (smart metering)
Interface B	Manufacturer interface
Internal CEM interface (optional)	Interface S2
Interface A	Interface S1

Fig. 1.3 Table showing compatibility between PAS 1878 functional components and the CENELEC/IEC architecture of TC205, TC59X and TC57

A wide variety of standards are available for the energy management of EVs in various architectural configurations [4]. The PAS maintains compatibility with these approaches as interface B and the manufacturer interface are allowed to be appliance (the Chargepoint) specific, and additionally, an external system interface is included in the architecture. The specification of these interfaces is not defined in PAS 1878, but they must meet basic functionality and security requirements to be PAS 1878 complaint. EV standards often contain various third-party operators; PAS 1878 covers these actors but splits out their constituent functional roles. For example, a charging station operator (CSO) in EV standards may be a combination of the

Chargepoint manufacturer and DSRSP functionalities in the PAS 1878 framework. The PAS deliberately divides the CSO functionality into these constituent parts in order to provide interoperability of DSR services. Whilst PAS 1878 defines these distinct functional roles, it does not specify who may (or may not) occupy the roles and whether multiple roles may be occupied simultaneously by the same actor; this may be a decision left to other standards implementations or national authorities.

1.2.1.2 Key Requirements

The PAS 1878 architecture is shown in Fig. 1.4, the full range of functional components are shown and whether they are mandatory or optional is indicated. In addition to the overall explanation accompanying Fig. 1.1, the following requirements and allowances underpin the architecture whilst facilitating versatile deployment:

- An ESA (the Chargepoint) must be supplied with a CEM (to deliver interoperable interface A), but may subsequently switch to another CEM.
- An ESA must only be connected to one CEM and one DSRSP at any one time (to avoid conflicting commands).
- A CEM may optionally be connected to multiple ESAs, provided each ESA is only connected to one CEM and one DSRSP (as above).
- A CEM must connect to only one DSRSP at any given time (to avoid conflicting commands).
- Multiple DSRSPs are allowed per premises; however, only one DSRSP must be connected to an ESA or CEM at any given time (as above).
- A HEMS may optionally be connected to multiple ESAs via their CEMs, and HEMS connectivity could be made interoperable by standardising interface C. (Note: Interface C functionality is not specified in PAS 1878; however, security requirements are.)
- CEMs and ESAs may optionally connect to external systems or a smart metering system.
- For an ESA which is a Chargepoint, an additional connection to an EV is required, this is a specific instantiation of an external system and external system interface.

1.2.2 Descriptions of Functional Devices and Entities

1.2.2.1 DSR Service Provider (DSRSP)

The DSRSP is responsible for managing DSR energy flexibility amongst its subscribed ESA portfolio (including Chargepoints) in response to requests sent by SOMPs. These SOMPs include transmission system operators (TSOs), distribution system operators (DSOs) and balance responsible parties (BRPs).

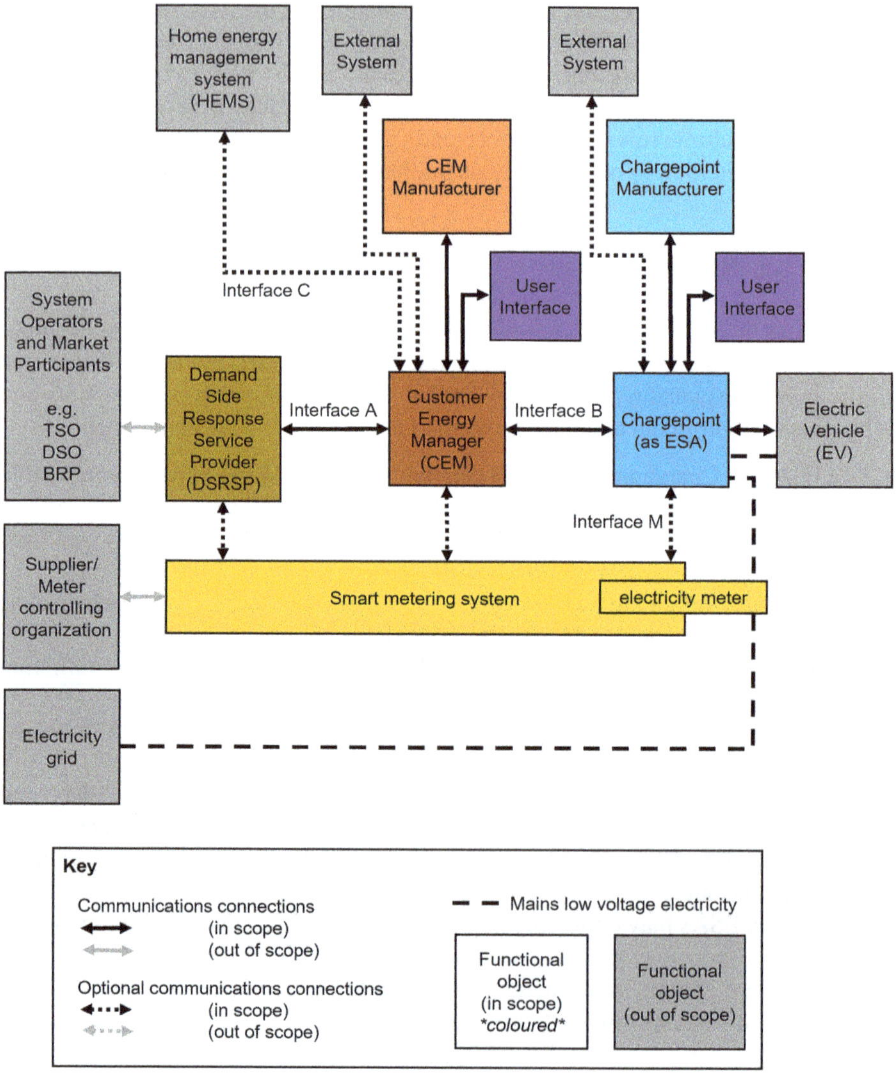

Fig. 1.4 Architecture of EV smart charging logical functionalities for DSR, showing all functionalities

The DSRSP receives sets of flexibility offers from each of the active ESAs in its portfolio via their associated CEMs. This allows the DSRSP to develop a view of the nature of flexibility options available to it over a given period and to match that with any incoming request from SOMPs.

When implementing a DSR event, triggered by a request from a SOMP, the DSRSP selects one of the flexibility offers received from each of its participating

ESAs (including Chargepoints) and sends this selection to each of the appropriate ESAs via their associated CEMs.

The DSRSP role may be performed by any of a range of organisations, perhaps working together, for example, an aggregator working in conjunction with a CSO. The National Electricity Regulator may optionally decide that only approved DSRSPs are allowed to operate, in which case DSRSPs should seek such approval.

The SOMPs and DSRSPs, and their relationships with ESAs, are described in more detail in PAS 1879.

1.2.2.2 Customer Energy Manager (CEM)

The CEM is included in several standardisation documents, including [12–14], where its functionality is described but not closely specified. PAS 1878 provides a baseline specification (a minimum set of requirements and behaviours) for the CEM in order to achieve interoperable communication and operation between the DSRSP and ESA within the context of DSR. For PAS 1878, compliance with this baseline specification is mandatory, and any other functionality included by the CEM manufacturer is allowed but not specified.

In addition to technical requirements, PAS 1878 also requires that each ESA (including Chargepoints) must be provided with its own (PAS compliant) CEM at the time of purchase, in order to guarantee that any ESA is able to connect to any DSRSP. This CEM may subsequently be replaced with another (PAS compliant) CEM of the consumer's choosing (providing the CEM and ESA are compatible).

The CEM acts as a logical intermediary between a DSRSP and an ESA, sending flexibility status information to the DSRSP and initiating ESA flexibility actions. It provides data model and communications protocol transcoding between those used on interface B and interface A (see Sects. 1.2.3.1 and 1.2.3.2), thus allowing semantic interoperability between the ESA and DSRSP. The requirement to conform to interface A on the "grid side" ensures that any (PAS compliant) CEM will be able to communicate with any (PAS compliant) DSRSP.

The CEM is seen as a logical function and may be implemented either as part of an ESA (including a Chargepoint) or in another device in the home or in the cloud. As the baseline functionality of the CEM requires a connection with only a single ESA and as there may be more than one CEM (and associated ESA) present in the premises, then the CEM may optionally connect to a HEMS over interface C. Although this interface is not specified in detail by PAS 1878, it does allow for the energy management of a range of ESAs within a premises in the case that all or some of them are individually connected to a CEM.

A securely connected local or remote user interface is used to input user preferences and DSR-related control and to display DSR status.

There are several device management and cyber security-related requirements placed upon the CEM by PAS 1878, including that CEM must be able to securely connect to a CEM manufacturer in order to provide software updates (including cyber security-related updates).

1.2.2.3 Home Energy Management System (HEMS)

A HEMS may connect to various electricity consumption, production and storage equipment within the premises, allowing them to be controlled according to rules and user preferences – e.g. selecting whether or not the energy from a photovoltaic system is (a) diverted to a Chargepoint in order to charge an EV or (b) exported to the grid, depending upon current and upcoming import/export tariffs and user requirements ("I need 100 miles charge 2 hours from now").

HEMS capability is particularly applicable to routine mode ESA operation, as it can be used to optimise ESA operation against various energy system incentives from within and outside the premises.

HEMS services will be provided by a HEMS service provider functionality, which is responsible for providing and operating any HEMS equipment and is contracted (directly or indirectly) to deliver HEMS services to the consumer. Which actors may deliver this functionality (possibly in addition to other functionalities) is not defined.

1.2.2.4 Chargepoint (The ESA Functionality)

Chargepoints are included in the list of equipment types that are eligible to be termed energy smart appliances (ESA). A Chargepoint is considered to be an energy smart appliance if it meets the requirements for ESAs in general and Chargepoints in particular, as described in PAS 1878 (Clause 8.1 specifies EV-specific requirements).

The Chargepoint determines and generates flexibility offers based on operating capabilities, consumer preferences (including expected departure/arrival schedule and journey distance/duration/characteristics) and external information (including EV battery capacity, EV state of charge and EV physical operating limits). If a Chargepoint supports vehicle-to-grid (V2G) functionality, which is non-mandatory, then it is also capable of determining and generating "production" flexibility offers, as described in Sect. 1.3.2.1.

A Chargepoint communicates with its CEM over interface B according to requirements described in Sect. 1.2.3.2. Registration, de-registration, authentication, flexibility offer, flexibility offer cancellation and status messages, to be shared with the DSRSP, are passed over interface B, and as this is critical information, cyber security requirements are specified.

The Chargepoint supports a secure connection over a different interface to its Chargepoint manufacturer in order to provide software updates (including cyber security-related updates).

A securely connected local or remote user interface is used to input user preferences and DSR-related control and to display DSR status.

Other connectivity and information exchange is permissible providing the communication meets the security requirements of PAS 1878. However, such functionality is beyond the scope of PAS 1878.

1.2.2.5 Chargepoint Manufacturer

The Chargepoint manufacturer is responsible for manufacturing the Chargepoint equipment and retains ongoing liability for the general operation of the equipment. The Chargepoint manufacturer provides and is responsible for the built-in DSR-based functionality of the Chargepoint, which includes the Chargepoint having the intelligence to create flexibility options and control operation in routine mode, e.g. a Chargepoint optimising the charging schedule against a TOUT to reduce costs for the consumer and creating other flexibility options to send to the DSRSP. The Chargepoint manufacturer also provides and is responsible for any non-DSR-related remote control of the Chargepoint, e.g. consumers setting charging schedules from a smartphone app purely for consumer preferences (and not considering any electricity system incentives).

The Chargepoint manufacturer ensures that the Chargepoint meets all the applicable requirements in PAS 1878, including user interface provision (via Chargepoint or associated CEM), software and firmware updates and device certificate management.

The Chargepoint manufacturer is responsible for supplying an individual (PAS complaint) CEM with each individual Chargepoint and is responsible for any CEM manufacturers operating on behalf of the Chargepoint manufacturer.

Note a manufacturer may contract third-party service providers to deliver some elements of functionality or operation that the manufacturer is responsible for providing; in these circumstances the manufacturer retains responsibility for any third parties operating on their behalf.

1.2.2.6 Electric Vehicle (EV)

The EV connects to the Chargepoint using either a proprietary or standard protocol [4]. Only standard protocols – e.g. ISO 15118-20 – are considered here, as discussed in Sects. 1.2.3.7 and 1.4.1. The EV may provide the Chargepoint with information that the Chargepoint requires to produce DSR flexibility offers, as described in Sect. 1.3.2.3.

The EV will have an associated EV manufacturer that is responsible for manufacturing the EV equipment and retains ongoing liability for the general operation of the equipment.

1.2.2.7 System Operators and Market Participants (SOMPs)

The following roles are SOMPs:

- *Transmission System Operator (TSO)*

  The TSO is responsible for operating the national electricity transmission network. The TSO procures DSR services for system balancing of supply and demand and for ancillary services to maintain quality standards for specific network characteristics.

- *Distribution System Operator (DSO)*
  The DSO is responsible for operating the local electricity distribution network. The DSO procures local DSR services for network management to maintain quality standards for specific network characteristics.
- *Balance Responsible Party (BRP)*
  BRP is a generic term for grid-side actors participating in wholesale markets; it covers suppliers, generators and other nationally specific entities, e.g. virtual lead parties in GB. A BRP procures DSR services to improve their imbalance position in the wholesale market, after gate closure.

When utilising DSRSPs, the SOMPs should follow national guidance on imbalances, should only use approved DSRSPs (if DSRSP approval is required by the Electricity Market Regulator) and should only use DSRSPs operating a trusted PKI architecture. The SOMPs should clearly specify for the DSRSP the technical parameters of the DSR service they require, including timing, quantity and location information.

1.2.2.8 Electricity Supplier

The electricity supplier is responsible for providing the supply of electricity to domestic consumers and for billing import and export of electricity for settlement.

The electricity supplier may facilitate slower timescale DSR services by setting import tariffs (e.g. TOUT) and export tariffs (e.g. feed-in tariff) which a Chargepoint can use as an incentive for optimisation in routine mode. These are typically set 24 hours in advance and in increments of 30 minutes to 1 hour, synchronised to auction slots in the national electricity markets.

Electricity suppliers and DSRSPs should not operate together in a manner which prevents ESA-DSRSP interoperability.

Note: An electricity supplier may also be a BRP, procuring DSR services as outlined above, and may itself take the role of DSRSP, depending on any national regulatory requirements.

1.2.2.9 National Electricity Regulator

The National Electricity Regulator sets regulations for participation in electricity markets and the operation of the electricity system.

Inappropriate operation of ESAs or DSR services could present a risk to grid stability. Therefore, the Electricity Market Regulator may optionally decide that only approved DSRSPs are allowed to operate. The details of any approval process, assessment criteria and administering organisation would be determined by the Electricity Market Regulator and should be proportionate to the risks presented.

1.2.3 *Descriptions of Interfaces*

1.2.3.1 Interface A

The DSRSP and the CEM exchange information relating to Chargepoint and CEM registration, de-registration, Chargepoint flexibility offers, DSR events, status and any cyber security incidents across interface A.

PAS 1878 defines mandatory information and messaging models which can be applied to lower-level data models and communications protocols.

In order to allow any Chargepoint, via its associated CEM, to be operated for DSR services by any DSRSP, a mandatory minimum implementation of interface A over OpenADR is defined in PAS 1878 (specified in Clause 5 and Annex F). Other additional implementations of interface A (using the information and messaging models) are allowed and are out of scope of PAS 1878. Any implementation of interface A must comply with a set of security requirements based around mutual authentication and encryption over TCP/IP using Transport Layer Security (TLS). The following provides a high-level summary of the interface A information model. Details may be found in PAS 1878:

- CEM and Chargepoint device information including manufacturer, serial number and firmware version
- Chargepoint power reporting capability and any DSRSP preference for capability
- Flexibility offers from Chargepoint to DSRSP (updated whenever Chargepoint status changes)
- Flexibility offer request (selection) from DSRSP to Chargepoint
- Acknowledgement of flexibility offer request from Chargepoint to DSRSP
- Actual power values (during a DSR event) from Chargepoint to DSRSP
- Cancellation of current DSR event (i.e. cancellation of "response mode" operation)

1.2.3.2 Interface B

The Chargepoint and CEM are connected using interface B, which must support the interface A information model. The data model and underlying protocols for interface B are not specified in PAS 1878, but it must be possible for the CEM to translate between the data models used on both interface A and interface B with no loss or corruption of information.

The same cyber security requirements are applied to interface B as are applied to interface A.

Example candidate open protocols for use over interface B include, but are not limited to:

- EN 16836-3 (Zigbee SEP 1.2)
- IEEE 2030.5 (Zigbee SEP 2.0)
- EN 50631 series (SPINE)

- IEC 62056 series (DLMS/COSEM)
- EN50491-12 series (interface between the home and building CEM)
- IEC 63110 (Protocol for management of electric vehicles charging and discharging infrastructures)
- EN 50090 series (KNX)

1.2.3.3 Manufacturer Interface

Both the CEM and the Chargepoint must support a logical interface to a manufacturer portal. The mandatory function of this interface is to provide firmware updates and certificate management information (new certificates, certificate revocation, etc.) to the CEM and Chargepoint during the authentication phase and CEM or Chargepoint de-registration notification during normal operation. The cyber security requirements for this interface follow those of interface A.

Whilst not defined in PAS 1878, example candidate open protocols for use over the manufacturer interface to a Chargepoint include, but are not limited to, IEC 63110 [4]. Section 1.4.1 contains further discussions.

1.2.3.4 Interface C

The optional interface C allows connection between a HEMS and one or more CEMs in order to provide enhanced energy management functions. Although this interface must follow the same cyber security requirements as interface A, the interface itself is out of scope of PAS 1878.

1.2.3.5 Interface M

The DSR system may interface to smart metering systems in order to gather information during routine mode, e.g. TOUT, or receive DSR requests from the DSRSP during response mode.

As the nature of smart metering systems varies from country to country, it is not feasible to include detailed interface specifications in the main body of PAS 1878.

However, possible methods for interfacing between the DSR system and the GB smart metering system are described in PAS 1878 (specification in Annex D) as an example. In the GB system, tariff information may be transferred to the Chargepoint (or CEM) using either the smart metering home area network (HAN), in which case the ESA (or CEM) needs to support Zigbee SEP 1.2, or via an external connection to the smart metering wide area network (WAN), in which case the ESA (or CEM) requires a connection to the Internet. In these cases, interface A or interface B may need to additionally support tariff information. Flexibility offers and flexibility offer requests may be transferred over the smart metering HAN and DCC WAN using an Auxiliary Proportional Controller (APC) attached to, or incorporated into, the Chargepoint. In this case, the DSRSP is a DCC user in the role of "import supplier".

1.2.3.6 External System Interface

The CEM and Chargepoint can connect to one or more external systems (e.g. weather service, grid carbon intensity monitoring service, smart metering service) in order to provide additional functionality. When this functionality is related to DSR and energy management functionality (directly or indirectly), the cyber security requirements for this interface follow those of interface A.

1.2.3.7 Chargepoint and EV Interface

This interface is used to provide the Chargepoint with information about the EV for the Chargepoint to create flexibility offers. This information includes the desired time of EV departure/arrival, expected range/journey characteristics, current battery state of charge, battery capacity, EV/battery operating limits and any expected V2G operation (if applicable).

This is an instantiation of an external system interface for the Chargepoint, so the cyber security requirements for this interface follow those of interface A.

Whilst not defined in PAS 1878, example candidate open protocols for use over the Chargepoint and EV interface include, but are not limited to, ISO 15118-20 as this protocol, when published, is likely to support security requirements and the communication of energy flexibility information [4]. Sections 1.3.2.3 and 1.4.1 contain further discussions.

1.3 Operation Framework

1.3.1 Operation Process and DSR Modes

The operation of the DSR system moves between phases (a) to (e) set out below.

Note: The term "Chargepoint" is used to mean an ESA which is an EV chargepoint.

1.3.1.1 (a) Consumer Registration with the DSRSP

In this phase, the consumer sets up an account with a DSRSP (possibly via a registration agent acting on behalf of the DSRSP) and may be provided with information used in subsequent phases (including authentication) depending upon the DSRSP business process (which is out of scope of PAS 1878).

1.3.1.2 (b) Discovery, Authentication and Device Registration

The Chargepoint and CEM discover and mutually authenticate each other, after which certain equipment information is transferred from the Chargepoint to the CEM. The CEM and DSRSP then mutually authenticate after which the Chargepoint and CEM are registered with the DSRSP (using the equipment information shared following Chargepoint/CEM authentication).

1.3.1.3 (c) Initialisation

Following a successful authentication and registration phase, the Chargepoint and DSRSP share information, via the CEM, about their operating capabilities during the initialisation phase. This information includes the actual power reporting types supported by the Chargepoint and the DSRSP's preferred type. Optional information, such as Chargepoint power consumption and/or production, may also be exchanged during this phase.

1.3.1.4 (d) Normal Operation

Normal operation phase includes the DSR reporting and requesting cycles. During this phase, the Chargepoint sends flexibility offers to the DSRSP, via the CEM. Whenever the Chargepoint status changes (e.g. the EV is plugged in or out), then it sends an updated set of flexibility offers to the DSRSP. The DSRSP receives flexibility offers from its portfolio of operating ESAs (including Chargepoints) and continually updates its flexibility capability forecast based on this information.

When a flexibility request comes to the DSRSP from a SOMP, then the DSRSP considers the parameters of the request and determines which ESAs to contact, via their respective CEMs, and which ESA flexibility offers to request.

Whenever it receives a flexibility request, the Chargepoint signals its acceptance (or not) to the DSRSP and begins execution of the requested flexibility offer. If the status of the Chargepoint changes during the execution of the flexibility offer (during the DSR event, e.g. the consumer presses an override button), then it provides the DSRSP with an updated set of flexibility offers. The DSRSP will then decide to newly request an updated flexibility offer or to exclude the Chargepoint from the current DSR event.

In order to allow smooth transitions into and out of DSR events, manage status changes due to ad hoc user interaction and deal with possible safety issues, both the CEM and Chargepoint are able to modify their behaviour when in the normal operation phase by moving between four "operation modes".

Mode 1: Routine Mode

- In routine mode (Mode 1), the CEM performs energy management functions in order to manage the consumption, production or storage of its connected ESAs (including Chargepoints). This management occurs according to consumer preferences and optionally other factors such as energy system incentives (e.g. TOUT), (pre-programmed) operating schedules, EV range/drive time targets, historic consumer behaviour, carbon intensity or weather forecasts. In order to enhance energy management within the premises, the CEM may be able to connect to a HEMS using interface C. The details of this HEMS functionality are outside the scope of PAS 1878.

Mode 2: Response Mode

- Once a Chargepoint accepts a flexibility offer request (by sending an acknowledge message to the DSRSP via the CEM), then it enters response mode (Mode 2). The Chargepoint will remain in response mode until either the flexibility offer period expires, the optional DSRSP-set duration expires, the DSRSP sends a cancellation request or the Chargepoint enters consumer override mode or failsafe mode. A Chargepoint will prioritise response mode operation above any routine mode operation.
- A CEM transitions its energy management functionality, including that involving a HEMS, from routine to response mode for any of its ESAs (including Chargepoints) that have entered response mode. If the CEM is also connected to ESAs (including Chargepoints) that are still in routine mode, then it is able to continue to manage them as routine mode ESAs.

Mode 3: Consumer Override Mode

- Whilst in Mode 1 or Mode 2, whenever a Chargepoint receives an ad hoc intervention from the user, then it transitions to consumer override mode (Mode 3) and operates to meet the ad hoc consumer wishes. User interventions include the modification of a current flexibility offer (when in Mode 2), rejection of a requested operation (when in Mode 1 or Mode 2), cancellation of an ongoing operation (when in Mode 1 or Mode 2) and cancellation of an ongoing operation for a specific period (when in Mode 1 or Mode 2).
- If the user cancels any operation when the Chargepoint is in Mode 2, then the Chargepoint transitions back to Mode 1.

Mode 4: Failsafe Mode

- Neither the DSRSP nor the CEM should request the Chargepoint to perform operations that are unsafe, harmful or otherwise hazardous. If any such requests are received by a Chargepoint, then its control logic should reject them. However, if it is clear that such conditions are imminent, then the CEM and/or Chargepoint transition into the failsafe mode (Mode 4).

- The details of this transition and the final condition are beyond the scope of PAS 1878 and are determined by the manufacturer.
- When transitioning into Mode 4, then the Chargepoint will always try to cancel any currently ongoing operations (in Mode 1 or Mode 2) and inform its CEM that it is entering Mode 4. The Chargepoint only exits Mode 4 when it is safe to do so.

1.3.1.5 (e) De-registration

De-registration of the Chargepoint, CEM and DSRSP may be initiated by the user. De-registration of a CEM may be initiated by the DSRSP.

In the case of user-initiated Chargepoint de-registration, then the Chargepoint securely removes all information related to its CEM and DSRSP. The CEM and DSRSP remove all information related to the Chargepoint. All communication is encrypted, all operations logged and the user informed of each main stage.

In the case of user-initiated CEM de-registration from the DSRSP, then the DSRSP removes all information related to the CEM and notifies the CEM. The CEM will then remove all information related to the DSRSP and inform the Chargepoint that it has been de-registered from the DSRSP. All communication is encrypted, all operations logged and the user informed of each main stage.

In the case that the DSRSP initiates CEM de-registration, then once the DSRSP receives a de-registration request from the CEM, the DSRSP will request confirmation from the user (consumer) before proceeding. Once confirmation is received, then the DSRSP will request the CEM to remove all information relating to the DSRSP. Once this is complete, then the DSRSP will securely remove all information related to the CEM. All communication is encrypted, all operations logged and the user informed of each main stage.

1.3.2 Power Profiles for DSR

1.3.2.1 Flexibility Offers as Power Profiles

The “flexibility offer” produced by a Chargepoint consists of a set of forecast power profiles and associated frequency response capability indicators.

The Chargepoint forecast power profiles provide the DSRSP with the possible options for Chargepoint power usage over a period of time and take the form of a set of power values at discrete, and possibly non-uniform, time intervals.

The Chargepoint has to provide a minimum set of forecast power profiles. In addition to this minimum set, which are given special labels, it may provide any

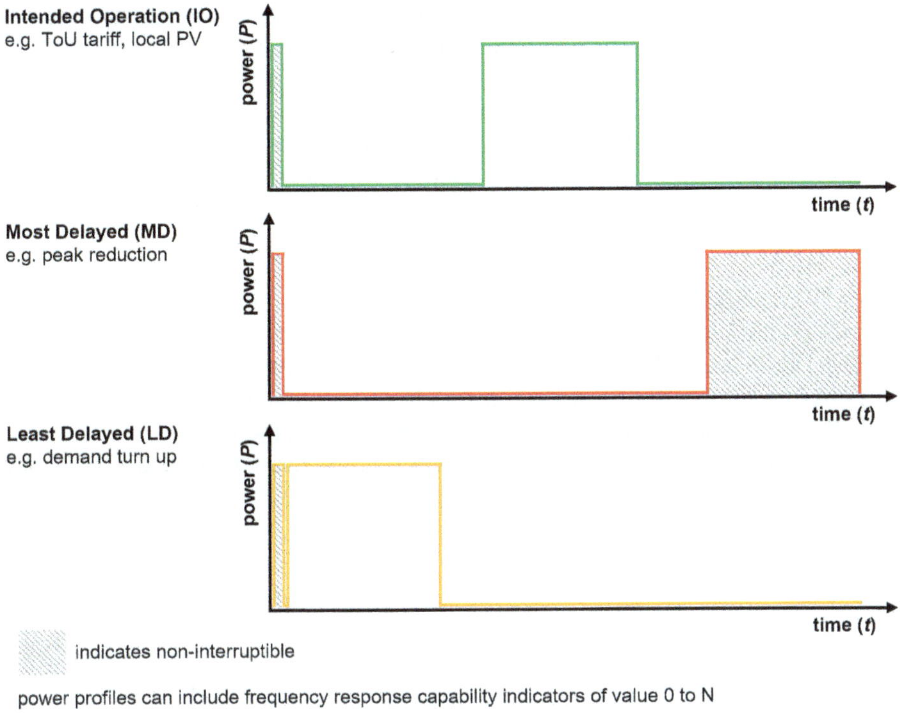

Fig. 1.5 Examples of the minimum three power profiles required for consumption only (non-V2G) chargepoints

number of additional forecast power profiles up to a combined maximum of 1000 profiles.

Power profiles indicate consumption of electricity with positive power values and production of electricity with negative power values. Power profiles indicate when operation cannot be interrupted (e.g. by a DSRSP request) by associating a non-interruptible indicator with power values (this is useful for washing machines with high rpm spin cycles which cannot be interrupted without safety consequences).

The power profiles are updated to the DSRSP whenever their status changes. This occurs at the start and end of any "active period" (defined as an interval within the power profile where the power is non-zero) and due to any consumer intervention or external input update.

Any Chargepoint acting as a load only has to produce three forecast power profiles as a minimum. These are shown in Fig. 1.5 and described below.

Intended Operation (IO)

- This profile is used for operation in routine mode.
- The "intended operation" (IO) power profile corresponds to the forecast power profile of the Chargepoint when not responding to a flexibility offer request. This is the profile that the Chargepoint expects to follow if it does accept a flexibility

offer request. This profile is generated by the Chargepoint within the context of consumer preferences and optionally any energy-related incentives such as TOUT or CO2 intensity.
- When implementing an IO profile, the Chargepoint must apply a randomised start time delay, up to 1800s (600s in Great Britain), unless a randomised start time is already incorporated due to being present in the incentive data. This randomised delay ensures that changes in aggregated load are slow enough for the electricity grid to cope with the changes in power.

Most Delayed (MD)
- This profile is an option for operation in response mode.
- The "most delayed" (MD) forecast power profile corresponds to the forecast power profile for the latest time at which a Chargepoint is able to start whilst still meeting the requirements of the user in providing the service.
- Most delayed forecast power profiles are not subject to randomised delays as they could be used to provide fast responding DSR services.

Least Delayed (LD)
- This profile is an option for operation in response mode.
- The "least delayed" (LD) forecast power profile corresponds to the forecast power profile for the earliest time at which a Chargepoint is able to start whilst still meeting the requirements of the user in providing the service.
- Least delayed forecast power profiles are not subject to randomised delays as they could be used to provide fast responding DSR services.

Most Delayed Production (MD_P) and Least Delayed Production (LD_P)
- These profiles are options for operation in response mode.
- Those Chargepoints that are able to produce power (e.g. storage batteries or (optional) EV V2G) generate two additional forecast power profiles – "most delayed production" (MD_P) and "least delayed production" (LD_P).

1.3.2.2 Frequency Response Indicator

When the demand on the electricity network increases and supply cannot increase fast enough to compensate, then the mains frequency drops below the desired value. Conversely when the demand on the electricity network decreases and supply cannot decrease fast enough to compensate, then the mains frequency rises above the desired value.

Some Chargepoints are able to monitor the mains frequency and adjust their consumption or production in order to move the mains frequency towards the desired value. A Chargepoint includes an indicator value associated with a power profile within a flexibility offer to indicate whether or not it is able to provide frequency response support and – if so – the degree of proportionality of the response with respect to frequency. The indicator is set from 0 (no frequency response capability) to N (dynamic frequency response capability with a response proportional to $(N-1)$th power of the frequency deviation).

The DSRSP indicates whether or not it would like the Chargepoint to enter "frequency response mode" within the flexibility offer request message. Two parameters are passed from the DSRSP to the Chargepoint via the CEM to allow the Chargepoint to modulate its power with frequency: the desired frequency values and the requested response sensitivity indicator. This allows a DSRSP to request the Chargepoint to achieve different levels of control, so giving a further tool to promote grid stability. When a Chargepoint is active in "frequency response mode", then it will begin to increase or decrease consumption or production (as appropriate) when the mains frequency deviates from the indicated values.

1.3.2.3 Information Required for Power Profiles

Consumer preferences must be obtained from the Chargepoint user interface or CEM user interface.

In order to create a set of flexibility offers, the Chargepoint needs such information as desired time of EV departure/arrival, expected range/journey characteristics, current battery state of charge, battery capacity, EV/battery operating limits and any expected V2G operation (if applicable).

The Chargepoint is able to obtain this information by several routes; all are external systems and would be accessed via the external system interface:

- Chargepoint – EV (EV passes the required information to Chargepoint over standard interface, e.g. ISO 15118-20)
- Chargepoint – EV manufacturer (EV provides the required information to EV manufacturer, and EV manufacturer passes the required information to Chargepoint over standard interface, e.g. IEC 63110)

For routine mode (Mode 1) operation, various energy system incentives or external data may optionally be obtained from external systems via the external system interface. For example, TOUT, generation source (renewable, nuclear, etc.), weather data and grid CO2 intensity may be obtained from:

- Chargepoint – electricity supplier portal (secure Internet link to electricity supplier portal, using external system interface)
- Chargepoint – DSRSP – DCC (in GB assuming that Chargepoint – DSRSP supports TOUT).
- Chargepoint – smart metering HAN (in GB requires Chargepoint to support Zigbee SEP to obtain smart metering HAN TOUT messages)
- Chargepoint – external data source (other information, such as weather, could be obtained from a secure external Internet connection, using external system interface)

1.3.2.4 Power Reporting

In order to allow the DSRSP to validate the actual performance of a Chargepoint during a DSR event against that provided in the flexibility offer, then the Chargepoint provides actual power values to the DSRSP. The actual power values may be transferred to the DSRSP either as discrete values at given times during the DSR event or stored and transferred as an actual power profile upon completion of the DSR event.

The Chargepoint must be capable of one or both of these methods. The DSRSP indicates its preferred option during the initialisation phase.

1.3.3 Cyber Security Approach

Cyber security protection is an important aspect of the design of the DSR system, and PAS 1878 specifies various mechanisms and requirements with the aim of protecting against cyber security incidents and of detecting and containing incidents when they happen.
The key aspects are:

1. Mutual authentication of all interacting entities using a PKI, making use of TLS
2. TLS-based encryption on all communications links using a specified cipher suite (as a minimum)
3. Certificate management
4. Requirement for newly registered CEMs and ESAs (including Chargepoints) to check and download the latest firmware over authenticated and encrypted communication channels
5. Strong key generation principles
6. Secure storage areas
7. Use of security logs to track activity
8. Requirement for ESAs (including Chargepoints) to implement secure boot processes
9. Anomaly detection and security incident management processes
10. High-level requirements placed on design, manufacturer and supply chain

In addition, a CEM and ESA (including Chargepoints) need to be compliant with all normative parts of ETSI EN 303 645 [15].

1.4 EV Smart Charging for DSR Services

1.4.1 Mapping to IEC/ISO Standards for EVs

There are several proprietary and standardised EV smart charging architectures and protocols in use today. In order to support the implementation of open and interoperable smart charging systems, PAS 1878 and PAS 1879 encourage the use of international standards. This is a very dynamic area, and so in order to provide an illustrative example of the main DSR processes described in PAS 1878, this section considers how two emerging standards – IEC 63110-1 and IEC 63110-2 and ISO 15118-20 – may be used within the PAS 1878 and PAS 1879 framework.

Note that publicly available information sources ("IEC 63110 Management of EV charging/discharging infrastructure", Paul Bertrand (Convenor of IEC TC69 JWG11), and "Communication Protocols for Electric Vehicle Charging – Meet ISO 15118-20", V2G – both presented as part of the Newcastle University smart charging webinar series [16]) have been used, and so details may be subject to modification prior to the publication of the given standards.

An example architecture is shown in Fig. 1.6. Other communication paths may be possible between the elements shown in Fig. 1.6, but as these are not currently being developed in international standards, then they are not considered here.
In Fig. 1.6, the ESA is represented by an aggregate of three functional entities:

1. The EV supply equipment (EVSE) is used to supply electrical power to the EV and to communicate with the EV. In this example, ISO 15118-20 is used as the communications protocol.
2. The charging station controller (CSC) is able to control the behaviour of one or more EVSEs.
3. The Local Charging Station Management System (Local CSMS) interfaces with EVSE (via the CSC) and cloud-based CSMS (operated by the CSO) in this example using IEC 63110 as the communications protocol. It should be noted that in addition to its core specification, IEC 63110-2 allows the use of additional functionality or information if required. The Local CSMS is able to communicate with the CEM within the context of routine and response mode operations either with or without consultation with the CSO cloud CSMS.

As discussed in Sect. 1.2.1.1, the charging station operator (CSO) may be a combination of the Chargepoint manufacturer and DSRSP functionalities of the PAS 1878 framework. The PAS deliberately divides the CSO functionality into these constituent parts in order to provide interoperability of DSR services.

The Chargepoint communicates using the Local CSMS with its CEM over interface B. Although the nature of this interface is not defined in detail in PAS 1878, it must support the interface A information model (or support an information model that allows the CEM to comply with the interface A information model) and comply with certain security requirements. Example candidate open protocols for interface B are described in Sect. 1.2.3.2.

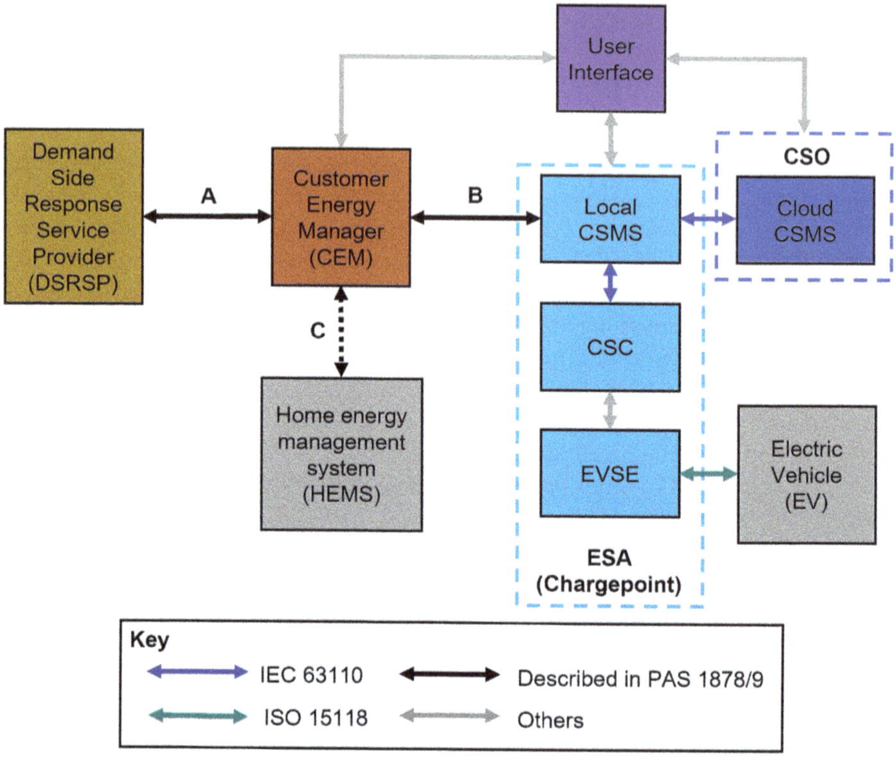

Fig. 1.6 Example of smart EV charger system integration in PAS 1878 framework

1.4.2 Example Use Case: EV Implementation for DSR Services

1.4.2.1 Registration

The first step in DSRSP, CEM and Chargepoint registration requires that both the CEM and Chargepoint perform any required software/firmware updates using their respective manufacturer interfaces. PAS 1878 mandates that this is performed using a mutually authenticated communications link using TLS v1.3 or above.

IEC 63110 specifies the use of XMPP at the transport layer. This supports TLS and certificate management and so provides a secure link between the Local CSMS (part of the Chargepoint) and the CSO (the Chargepoint manufacturer).

If the CEM is also provided by the CSO, then the same protocol and mechanisms could be used over interface B, thus also meeting PAS 1878 requirements.

The remaining steps in the discovery, authentication and initialisation phases are then performed as outlined in Sects. 1.3.1.1, 1.3.1.2, and 1.3.1.3.

The Local CSMS is able to obtain information from both the EVSE and the cloud CSMS in order to assist it in providing information on actual power reporting capabilities, etc. during this process.

1.4.2.2 Normal Operation

Mode 1: Routine Mode

When operating in routine mode, the Local CSMS (in the Chargepoint) may be able to obtain such information as EV presence, EV charging status, battery level, required charging targets/schedule/profiles and frequency response capability from the EVSE, the cloud CSMS and/or the EV itself (via the EVSE and CSC).

In addition, the Chargepoint may be able to obtain information such as TOUT or grid CO2 intensity from the local smart metering HAN, the Internet, the CEM or the CSO (if such information is available to the CSO).

The Local CSMS processes this information and schedules expected charging operations accordingly. The expected charging operation includes a power profile and so is used, along with the frequency response capability indicator, to form the "intended operation" power profile of the current DSR flexibility offer. Additional limiting factors are taken into account in order to produce the "least delayed" and "most delayed" power profiles. These are forwarded to the DSRSP via the CEM and updated whenever there is a change in Chargepoint status.

If the CEM is connected to a HEMS which is connected to other CEMs, then the Chargepoint may participate in wider energy management operations within the premises.

Mode 2: Response Mode

Whenever the Local CSMS receives a flexibility offer request from the DSRSP, via the CEM, then it informs the EVSE of the required power profile (or frequency response mode invocation). The EVSE will then provide power to the EV according to the chosen profile. The Local CSMS may also inform the CSO of the current DSR status and chosen flexibility offer, although this is beyond the scope of PAS 1878.

1.4.2.3 De-registration

When the Chargepoint owner (user) wishes to de-register the Chargepoint from either the CEM or the current DSRSP, then they must instigate this through the Chargepoint/CEM user interface.

As part of this process, the Chargepoint will communicate with the CEM over interface B, which will communicate with the DSRSP over interface A and will remove any information relating to the DSRSP or CEM (accordingly). Information relating to the Chargepoint will be removed from the CEM or DSRSP (accordingly).

If the Chargepoint is de-registered from its DSRSP but still maintains registration with its CEM, then it may be able to participate in routine mode operation (depending upon how TOUT and CO2 information, etc. are obtained) but will no longer be able to participate in response mode DSR operations.

When the Chargepoint is de-registered from its DSRSP by the DSRSP itself, then both parties remove information relating to the other.

Acknowledgements The concepts presented in this chapter are based on the technical framework set out in PAS 1878 and PAS 1879, which are free to download from the British Standards Institution (BSI) website [17]. BSI facilitated the Publicly Available Specification (PAS) development process, and the hard work of the whole BSI team was much appreciated. Many thanks are also extended to the UK industry experts, from across the energy, transport, built environment, communications and cyber security sectors, who sat on the Steering Groups leading the development of the PAS standards and who sat on the Strategic Advisory Group for the programme. The greatest thanks and acknowledgement go to the technical authors, Lee Gould and Hugh Boyes, who drafted PAS 1878 and 1879, respectively.

It is gratefully acknowledged that the UK Department for Business, Energy and Industrial Strategy (BEIS) funded the development of PAS 1878 and 1879. The PAS standards work supports actions set out in the Smart Systems and Flexibility Plan [18]. Other relevant policy documents include the BEIS Smart Appliances Consultation [19] and the Office for Zero Emission Vehicles (OZEV) Smart Charging Consultation [20].

Thanks are also extended to Dr Myriam Neaimeh at The Alan Turing Institute, who we are very grateful to for providing technical expertise on EV smart charging and associated standards.

Bibliography

1. BEIS. (2021). Modelling 2050 – Electricity system analysis. [Online]. Available: https://www.gov.uk/government/publications/modelling-2050-electricity-system-analysis
2. Hildermeier, J., Kolokathis, C., Rosenow, J., Hogan, M., Wiese, C., & Jahn, A. (2019). Smart EV charging: A global review of promising practices. *World Electric Vehicle Journal, 10*(4), 80.
3. Spencer, S. I., Fu, Z., Apostolaki-Iosifidou, E., & Lipman, T. E. (2021). Evaluating smart charging strategies using real-world data from optimized plugin electric vehicles. *Transportation Research Part D: Transport and Environment, 100*, 103023.
4. Neaimeh, M., & Andersen, P. B. (2020). Mind the gap- open communication protocols for vehicle grid integration. *Energy Informatics, 3*, 1.
5. ESC. Consumers, vehicles and energy integration. [Online]. Available: https://es.catapult.org.uk/case-study/consumers-vehicles-and-energy-integration/
6. BEIS. Beyond Off Street. [Online]. Available: https://www.gov.uk/government/publications/smart-meter-electric-vehicle-charging-competition-winning-projects
7. WPD. Electric Nation. [Online]. Available: https://electricnation.org.uk/resources/smart-charging-project/
8. UKPN. Shift. [Online]. Available: https://innovation.ukpowernetworks.co.uk/projects/shift/
9. SSEN. My Electric Avenue. [Online]. Available: https://www.ssen.co.uk/myelectricavenue/
10. BSI. (2021). PAS 1878:2021 Energy smart appliances – System functionality and architecture – Specification. [Online]. Available: https://www.bsigroup.com/en-GB/about-bsi/uk-national-standards-body/about-standards/Innovation/energy-smart-appliances-programme/pas-1878/

11. BSI. (2021). PAS 1879:2021 Energy smart appliances – Demand side response operation – Code of practice. [Online]. Available: https://www.bsigroup.com/en-GB/about-bsi/uk-national-standards-body/about-standards/Innovation/energy-smart-appliances-programme/pas-1879/
12. CENELEC. (2018). EN 50491-12-1:2018, General requirements for Home and Building Electronic Systems (HBES) and Building Automation and Control Systems (BACS) – Smart grid – Application specification – Interface and framework for customer. In *Part 12-1: Interface between the CEM and home/building resource manager – General requirements and architecture*.
13. CENELEC. (2017). EN50631: 2017, Household appliances network and grid connectivity. In *Part 1: General requirements, generic data modelling and neutral messages*.
14. IEC. (2015). TR 62746-2:2015, Systems interface between customer energy management system and the power management system. In *Part 2: Use cases and requirements*.
15. ETSI. (2020). EN 303 645 V2.1.1 (2020-06), CYBER; Cyber security for consumer internet of things. In *Baseline requirements*.
16. N. University. Smart charging webinars. [Online]. Available: https://www.ncl.ac.uk/cesi/events/webinars/v2gwebinars/
17. BSI. (2021). Energy Smart Appliances Programme. [Online]. Available: https://www.bsigroup.com/en-GB/about-bsi/uk-national-standards-body/about-standards/Innovation/energy-smart-appliances-programme/
18. BEIS. (2021). Transitioning to a net zero energy system: Smart systems and flexibility plan 2021. [Online]. Available: https://www.gov.uk/government/publications/transitioning-to-a-net-zero-energy-system-smart-systems-and-flexibility-plan-2021
19. BEIS. (2018). Government response to consultation on proposals regarding smart appliances. [Online]. Available: https://www.gov.uk/government/consultations/proposals-regarding-setting-standards-for-smart-appliances
20. BEIS. (2021). Electric vehicle smart charging: Final outcome. [Online]. Available: https://www.gov.uk/government/consultations/electric-vehicle-smart-charging

Chapter 2
The Concept of Li-Ion Battery Control Strategies to Improve Reliability in Electric Vehicle (EV) Applications

Foad H. Gandoman, Vahid Nasiriyan, Behnam Mohammadi-Ivatloo, and Davood Ahmadian

2.1 Introduction

In recent years, the development of electric vehicles (EVs) has exploded. Rechargeable batteries, which are utilized as energy storage devices, are becoming more important in EV applications [1]. Lithium-ion batteries, with their increased energy density, high energy efficiency, low weight, and long cycle life, may be an excellent choice for powering an electric vehicle in this situation. Typically, an EV with lithium-ion battery storage can only drive a certain distance at full power [2]. Because of the restricted useable range, a battery charging system and a different battery control strategy in EV applications are required.

Batteries are gradually gaining traction as one of the most critical components of future transportation systems. A Li-ion battery pack's battery management system (BMS) is one of the most important components. The BMS is responsible for controlling discharging and charging strategies, voltage monitoring, battery balance, temperature management in cells and battery packs, problem diagnostics, and system dependability. In general, there are three key performanceareas in the BMS

F. H. Gandoman (✉) · V. Nasiriyan
Faculty of Electrical and Computer Engineering, University of Tabriz, Tabriz, Iran

B. Mohammadi-Ivatloo
Electrical and Electronics Engineering Department, Muğla Sıtkı Koçman University, Muğla, Turkey
e-mail: mohammadi@ieee.org

D. Ahmadian
Faculty of Mathematics, Statistics and Computer Sciences, University of Tabriz, Tabriz, Iran
e-mail: d.ahmadian@tabrizu.ac.ir

V. Vahidinasab, B. Mohammadi-Ivatloo (eds.), *Electric Vehicle Integration via Smart Charging*, Green Energy and Technology,
https://doi.org/10.1007/978-3-031-05909-4_2

system: fundamental aspects, main hardware aspects, and main software aspects. The main aspects of BMS in the Li-ion battery pack are battery cell monitoring [3, 4]; input/output current and voltage monitoring [5, 6]; charge and discharge control [7, 8]; estimations such as the state of charge (SoC) [9, 10], the state of health (SoH) [11], and the state of function (SoF) [12]; battery protection [13, 14]; cell balancing and equalization [15]; temperature control and heat management [16]; communication and networking [10]; and fault diagnosis and assessment [12].

As Li-ion batteries age, they develop metallic lithium plating and the solid electrolyte interphase layer and lithium dendrite, which may lead to dangerous failures if they are operated under certain circumstances, such as overcharging, overdischarging, or running at a low temperature [17]. As a result, the existence of BMS fault detection and diagnostic functions is crucial for Li-ion battery-driven systems, particularly for high-power applications. A large number of research studies were carried out in order to develop an accurate, reliable, robust, and easy-to-apply fault diagnostic method. The importance of developing an efficient fault diagnostic system for the evolution of Li-ion battery-powered systems was studied in [18]. Reference [19] presents a comprehensive overview of the failure process of Li-ion batteries and potential remedies via state-of-the-art review research. An overview of alternative problem diagnostic techniques for Li-ion batteries was examined [20].

Estimations of the SoC, SoH, and open-voltage circuit (OCV) play an important role in improving the control strategy and lifetime of the Li-ion batteries [21]. The SoC estimate of Li-ion batteries tries to verify a battery's remaining capacity throughout a charge-discharge cycle, preventing the battery from overcharging and overdischarging [22]. The purpose of the SoH calculation is to anticipate the remaining usable life of charge-discharge cycles, which determines whether Li-ion batteries need to be replaced. The battery SoH, on the other hand, is characterized by slow-changing properties like capacity fading and resistance rising, and it fluctuates with cycles, necessitating long-term monitoring [23].

In [24], the virtues and drawbacks of Li-ion battery SoC estimates based on battery models, estimation techniques, and their merits and drawbacks were discussed. The erroneous causes influencing the SoC estimate were investigated in [25], but the review of the estimation technique was brief. Reference [26] just highlighted the fundamental problems and provided a summary of the live SoC estimate based on the Kalman filter (KF) family of algorithms. Rather than model-based SoC estimation, [10] focused on data-driven SoC estimation approaches for Li-ion batteries in EV applications. [27] divides battery SoH estimates between experimental and model-based techniques but does not go into detail about the merits and downsides of each estimating method.

This chapter presents a brief overview of the battery control strategy based on the BMS operation in Sect. 2.2. In Sect. 2.3, a case study based on the voltage variation of the battery at different ambient temperatures over time has been investigated. The conclusion has been explained in Sect. 2.4.

2.2 Battery Management System (BMS)

EVs have a BMS that controls and monitors an EV's battery system. The BMS maintains battery safety and dependability, improves efficiency, extends battery life, and boosts EV range. A battery system is essential since a single cell has limited capacity and voltage. A battery system is made up of several battery cells linked in series, parallel, or a mix of both. The BMS is responsible for estimating battery states and characteristics, preventing overcharging, overdischarging, thermal runaway, and battery inconsistencies [25]. In terms of hardware implementation, the BMSs have five primary tasks [28]: battery parameter acquisition, battery system balancing, battery information management, battery thermal management, and battery charge regulation. Some of the combined features of battery storage in the BMS are as follows:

- Expenditure: consists of producing, reparation, function, and transforming expenditure
- Life cycle: determined by charging or discharging
- Power supply: determined to charging or discharging amount, energy storage amount
- Security: determined to the security or hazardous elements because of functional temperature

The BMS has some tasks that are as follows:

- Cell monitoring: consists of current, voltage, and temperature situation.
- Battery safety: one feature of the operation of BMS is to provide the safety of the battery and eliminate it from dangerous situations.
- SoC evaluation: SoC is a metric of the existing charge storage in the battery that shows the entire status of battery charging.
- SoH evaluation: SoH is a significant metric of battery operation feature; SoH forecasts the quantity of charging and discharging duration.
- Cell regulation: consists of various arrangement battery cells to supply suitable functions to the battery system.

In general, there are four kinds of topology for the BMS in EVs. Table 2.1 describes the advantages and disadvantages of the topologies.

BMS may be simple or complicated, and they can use a variety of technologies to fulfill their primary goal, which is "taking care of the battery." On the other hand, these systems may be classified depending on their topology, which refers to how they are deployed and work throughout the battery pack's cells or modules. The BMS has some elements that are as follows:

- CAN-bus component which is applied for relating with control elements.
- DC/DC component is applied for a convert power source into the power supply.
- The battery current evaluation component is applied for calculating the charging and discharging current situation.

Table 2.1 Definition advantage and disadvantages of different BMS's topologies

Type	Definition	Advantages	Disadvantages
Centralized BMS	The battery pack has one central BMS, and all the battery packages are connected to the central BMS directly	Because there is just one BMS, it is more compact and tends to be the most cost-effective	Because all of the batteries are linked directly to the BMS, the BMS requires a large number of ports to connect with all of the battery packages
Modular BMS	The BMS is separated into numerous replicated modules, each with its dedicated bundle of wires, and connects to a battery stack's neighboring allocated section	Troubleshooting and maintenance are simplified because of the inherent flexibility, and expansion to bigger battery packs is simple	The disadvantage is that total expenses are somewhat greater, and depending on the program, there may be duplicated unnecessary functionality
Subordinate BMS	Similar to the modular architecture in concept, however in this instance, the slaves are limited to just relaying measurement data, while the master is devoted to calculation, control, and external communication	Because the slaves' functionality is often simpler, there are likely less overhead and fewer unwanted features; the expenses may be cheaper than with a modular BMS	The disadvantage is like the modular BMS
Distributed BMS	The electrical hardware and software are housed in modules that connect to the cells through bundles of wires. All of the electrical gear for a distributed BMS is housed on a control board that is mounted directly on the cell or module being monitored	The majority of the wiring is reduced to a few sensor wires and communication cables between nearby BMS modules. In addition, each BMS is more self-contained, allowing it to conduct calculations and communications as needed	Troubleshooting and maintenance might be difficult with this integrated form. Because there are more BMSs in the total battery pack system, costs are likewise greater

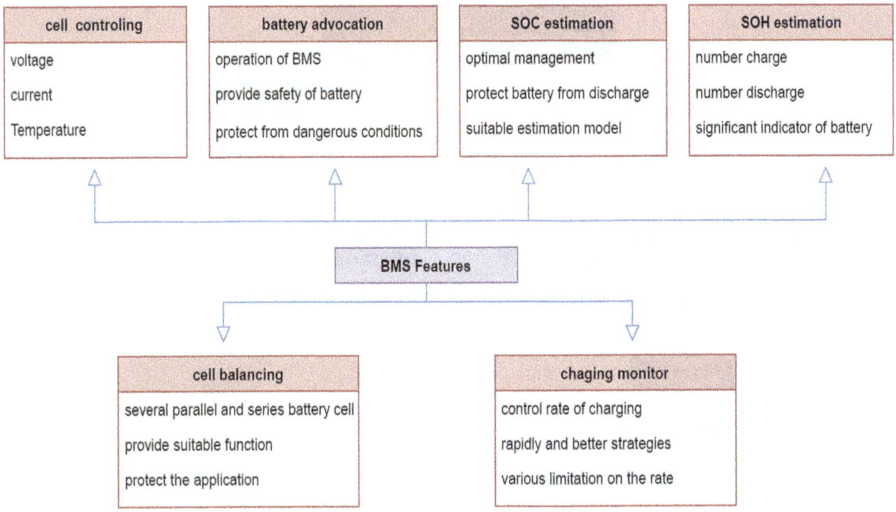

Fig. 2.1 BMS features configuration

- The battery voltage evaluation component is applied to calculate the voltage situation.
- The temperature evaluation component is applied to check the situation of selected cells.
- Data flash memory component is applied for recorded data of evaluation.

All components above must be related to the selected control element to operate properly. Figure 2.1 shows the full picture of the BMS features configuration.

2.3 Battery Fault Detection

It is one of the primary duties of a BMS to safeguard both the battery and its users from the dangers connected with its operation. Most dangerous circumstances arise from defects, and the BMS should be able to reduce both the probability and severity of these problems. In order to keep the battery system safe, several components such as sensors, contactors, and insulation are often included. Sensors attached to the cells keep tabs on the cells' voltage, current, and temperature restrictions. These safety precautions, however, are typically insufficient because of the complexity of the BMS's hardware and software implementation. As a result, BMS must have algorithms for diagnosing faults. It is the goal of these algorithms to identify flaws as early as possible and to provide the battery and its users with the required and quick management measures. Figure 2.2 shows the potential for internal and external faults in the Li-ion batteries.

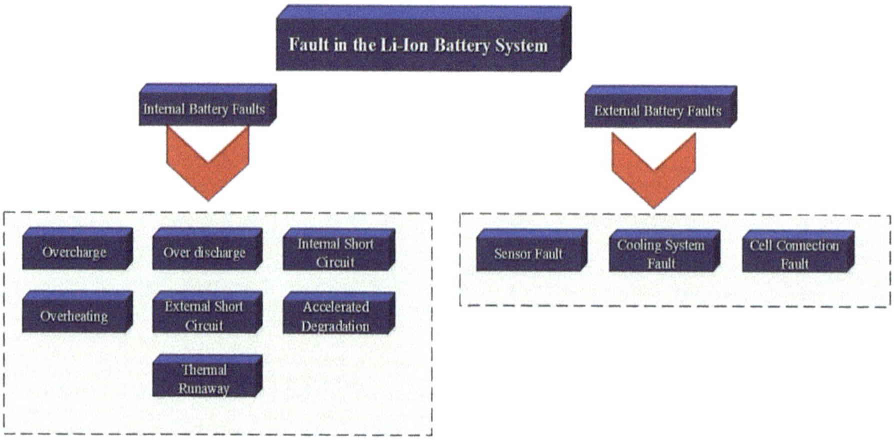

Fig. 2.2 Potential internal and external faults categories in the Li-ion batteries

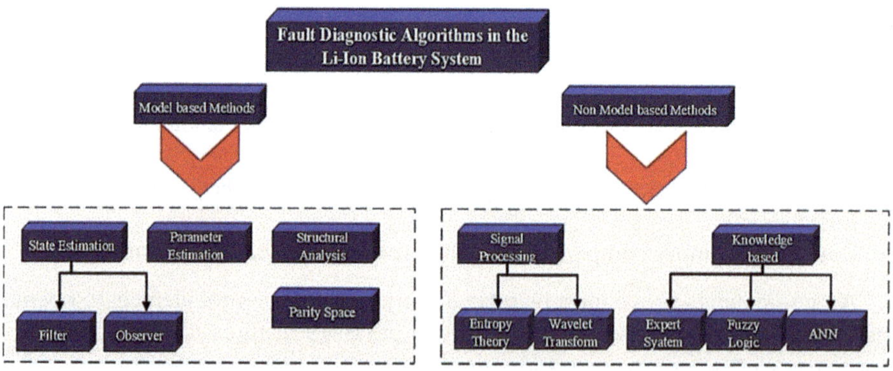

Fig. 2.3 Most recent advancements in algorithmic techniques for fault identification

The BMS is critical in failure diagnosis because it contains all the diagnostic subsystems and algorithms. It uses sensors and state estimates, as well as modeling and data analysis, to keep tabs on the battery system and alert the user if anything unusual occurs. It is difficult to perform this duty effectively because of several internal and external faults. Many fault diagnostic techniques must operate together in order to accurately identify and isolate a particular issue and apply the necessary control action. Because of this, the BMS's defect detection techniques are constrained in terms of both processing space and time. While preserving accuracy and reliability, these fault diagnosis algorithms must have minimal computing effort in certain applications because of the vast number of battery cells in the battery system. Thus, fault diagnosis techniques for Li-ion batteries have been the subject of substantial study and development in recent years. Figure 2.3 depicts the most recent advancements in algorithmic techniques for fault identification.

Model and signal processing-based approaches have been the most commonly used for detecting Li-ion battery system failures. However, the use of machine learning-based approaches is expanding at a much quicker rate due to some of the notable benefits, such as a high degree of accuracy, compatibility with the highly nonlinear Li-ion battery system, and decreased dependency on domain specialists. An accurate equivalent circuit model of the Li-ion battery is essential for model-based fault diagnosis procedures to be accurate and reliable. The intrinsic features of the highly nonlinear Li-ion battery remain a mystery, making it difficult to obtain an extremely precise model. With the introduction of machine learning-based approaches, this constraint is removed. As a result of the use of machine learning techniques, measurement noises that restrict the use of signal processing methods have greatly decreased. To further simplify fault diagnosis, machine learning-based techniques eliminate the need to collect the battery's physical information and to learn nonlinear correlations between battery internal parameters and external measurements such as operating current, the terminal voltage, and temperature. These ultimately minimize the need for domain-specific expertise, time, and the cost of developing the system itself.

The modeling techniques and data achieved techniques have extrapolated features that are simple and design features, respectively. For fault detection, the fault detection and diagnosis (FDD) algorithm has been used to adjust conditions in an intermediate or faulted situation. Figure 2.4 shows an overview of the FDD algorithm. The FDD algorithm applies to some of the applications that are as follows:

- Applying joint confidence region (JCR) profiles-model realization estimated by applying the JCR model. The FDD algorithm is applied to dissolve converse

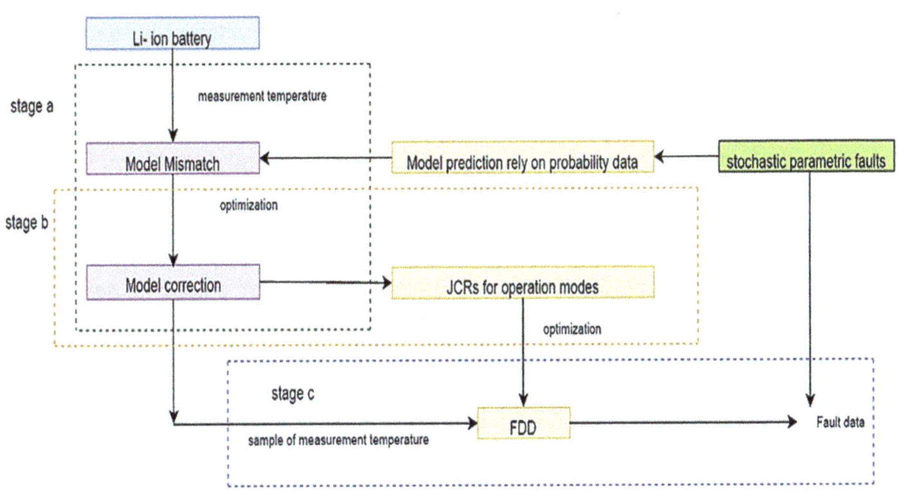

Fig. 2.4 Overview of FDD algorithm

difficulty and realization uncertainty by applying JCR models. Stage (a) the haphazard distribution of faults, applying a JCR profile of internal temperature for an average price of fault; stage (b) applying JCR profiles of internal temperature for average prices of faults; stage (c) applying a JCR model based on least interspace optimization for FDD applications due to JCR noise problems.
- The FDD algorithm applies JCR profiles and the computationally efficient average price of resistance by solving determined problems and supplying the contingency data of occurring faults.
- This application makes the assumption that the battery models are correct and employs model mismatch.

Battery safety has improved with the fault diagnosis techniques outlined, but there are still certain limits in real-life applications. However, great modeling accuracy is required for model-based solutions to identify and isolate a defect promptly and in real time. Greater knowledge of the internal battery activity is needed to construct an accurate, but not unduly complex, battery model. Additional considerations include the trade-off between the diagnostic approach's robustness and sensitivity, which is dependent on the fault thresholds. It is possible to bypass some of the more onerous constraints for battery modeling by using non-model-based solutions. The dynamic performance of signal processing algorithms is high, but they are susceptible to measurement noise and cannot consistently identify early defects. Detecting battery issues using expert systems, which rely on simple knowledge-based approaches, is difficult since expert systems need good rules to identify problems accurately. Although machine learning and other complex knowledge-based approaches work well with nonlinear systems like Li-ion batteries and are very accurate, the training process is lengthy and requires a huge quantity of data. Model-based solutions are becoming widespread, although there are still significant challenges in improving battery model accuracy, particularly throughout the full battery life. Approaches that are not reliant on models, like data-driven methods, may be very helpful in anticipating battery degradation and in the model creation process itself. Therefore, a mix of both model-based and non-model-based methodologies should be used for Li-ion battery problem diagnostics.

All of these approaches to fault diagnosis encounter several difficulties. It is hard to pinpoint the exact reason for a problem when there are so many possible causes with comparable impacts on the battery. To avoid having to isolate and identify the observed defects, current approaches generally presume that other components in the system are working properly. Further research should concentrate on developing new techniques for identifying faults after a successful detection of a flaw. Due to a lack of knowledge about fault behavior, determining appropriate fault thresholds for early and accurate detection is also a difficult challenge. In addition, developing fault simulation tools needs a deeper knowledge of battery fault behavior since creating a true physical failure may be impracticable, expensive, and dangerous. There is a need for further research on the behavior of defects. For modeling and simulation purposes, this must be done via well-designed experiments. Last but not least, the processing power of the BMS must be increased to handle additional data. Improved fault detection accuracy may be achieved via the use of complicated algorithms. The BMS and cloud-based battery status monitoring technologies may both help to attain this goal in the future.

2.4 Battery State-of-Function Estimation

SoH and SoC estimations are two key parameters in the Li-ion battery which play an important role in increasing the reliability and lifetime of the system. In this section, the potential method for the estimation of these two important parameters will be discussed.

2.4.1 Battery SoH Estimation

These three main methods have been used for the SoH estimation of the battery: experimental methods, model-based methods, and machine learning methods. Figure 2.5 shows the SoH estimation methods for the Li-ion battery system.

- Experimental methods: Since they typically need specialized equipment and take a significant amount of time, most experimental approaches are carried out in a formal laboratory setting. Measurements and data collection are the foundation of these approaches to better understand and assess battery aging.
- Model-based methods: A measure of battery degradation, the SoH is a measure of the battery's state of health (SoH). However, direct measurements cannot be used to calculate it. SoH indications must be made using the capacity of the battery, the resistance of the battery, or the impedance of the battery.
- Machine learning methods: Using a mix of experimental and model-based techniques, these methodologies may be used. The SoH of a battery may be estimated using training data, measurements, and models.

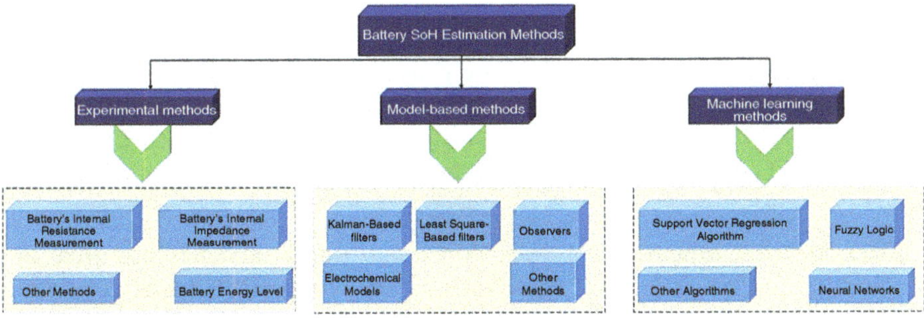

Fig. 2.5 SoH estimation methods for Li-ion battery system

Real-time BMS applications seem to make it harder to use methodologies based on experimentation. On the other hand, automobile applications need the real-time deployment of model-based and machine learning technologies. This is a basic structure with an accurate estimate and straightforward execution in both of these kinds of categories. Fully electric car and hybrid electric vehicle data, however, is not currently readily available for machine learning techniques. Furthermore, they might become unreliable when confronted with unexpected operational situations that were not part of their training [29]. The model-based SoH approaches, on the other hand, seem to be the most practicable. Adopting an adaptive filtering strategy is one of the model-based approaches that have been considered and have shown promising outcomes in terms of accuracy, simplicity, and computing practicality. Equivalent circuit models are used to identify the internal resistance of the battery.

In terms of battery capacity, this SoH indication is well within reach with the equivalent circuit models in hand. For hybrid applications, in which the battery is the primary source of power and its energy capacity is less important than its power, this signal is of considerable relevance in determining the battery's power capability based on SoH [30].

Adaptive filtering based on a Kalman algorithm is more suited for state estimation than the standard Kalman filter because of the battery's nonlinear nature. Support vector regression using equivalent circuit models, on the other hand, was the focus of numerous investigations owing to its simplicity and a high degree of precision. The precision and simplicity of this approach make it an excellent choice for estimating battery characteristics. Another advantage of this assessment is that one may estimate SoH by looking at the battery internal resistance [31, 32]. The equivalent circuit models in hybrid applications may be improved by updating all of the model parameters. A nice example of this onboard SoH estimate may be shown using the support vector regression technique, which is based on model-based estimation [33]. Due to its superior performance when compared to the other approaches, this adaptive filter was chosen to serve as an example of how battery SoH might be estimated in real time for automotive use.

2.4.2 Battery SoC Estimation

When estimating battery SoCs, there are generally four methods to choose from: (1) the direct measurement approach that is not reliant on the battery model, (2) the so-called "black box" model, (3) the so-called "state space" model, and (4) the so-called "hybrid" model. Figure 2.6 shows the SoC estimation methods for the Li-ion battery system.

1. The direct measurement approach: Since the battery model is not used, this technique estimates the SoC based on a variety of repeatable battery characteristic variables that have a strong relationship to the battery, such as the voltage, current, internal resistance, and impedance.

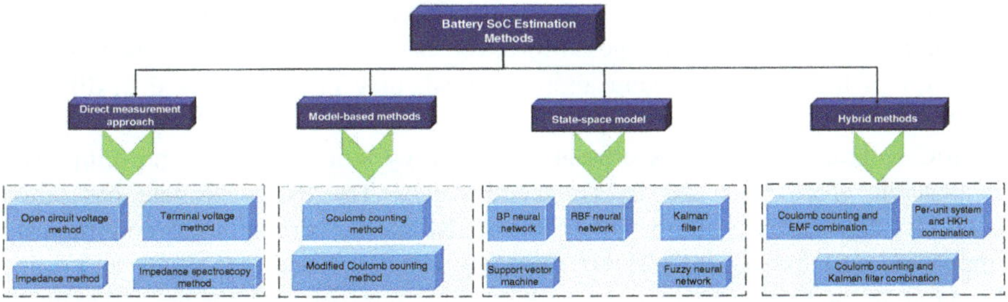

Fig. 2.6 SoC estimation methods for Li-ion battery system

2. The black box model: An unknown system is treated as an input to the black box battery model, which is based on the battery's online observable current, voltage, temperature, and so on as an output.
3. The state space model: An observer or observer filter is used to estimate a battery SoC using a model of the battery's power consumption as a state variable in the system state space expression.
4. Hybrid methods: The benefits of each SoC estimate approach are combined in hybrid models, resulting in a globally optimum estimation performance. When compared to individual approaches, the research suggests that hybrid methods offer a decent estimate of SoC.

The open-circuit voltage approach has the best estimation accuracy for direct detection methods that do not depend on the battery model, but it is difficult to apply to real vehicle operation since it needs a long time to stand. The ampere-hour integration technique is also commonly used because of its simplicity, dependability, and cheap equipment requirements. Although it is highly dependent on the accuracy of the starting value and the accuracy of the sensor measurement value, it may still be useful [36]. There will be accumulated mistakes over a lengthy period of integration. The estimated accuracy of the internal resistance, discharge test, and electrochemical impedance techniques is excellent, but the equipment and testing circumstances are strict. The load voltage approach is not suited for actual applications because of the circuit's influence. Using a black box battery model, the SoC estimate approach offers a high degree of learning ability and a high degree of accuracy [37].

2.5 Conclusions

This chapter discussed the control strategy of Li-ion batteries as the main component of EVs. Different battery management systems have been investigated. Moreover, different methods for estimating the key parameters of the battery, such as SoC and SoH, for a better understanding of the lifetime performance of the battery have been introduced. The proposed method for state-of-function estimation of the batteries has advantages and disadvantages. Moreover, based on the rate of sensitivity and accuracy of the data processing and command, they have been used in different electric mobility applications.

SoC estimate accuracy is crucial since energy storage devices are employed in portable electronics and hybrid electric cars. The SOC estimate has lately been investigated by a number of scholars. Estimation precision has continuously improved, and extensive R&D is currently ongoing. SoC estimates and some literature in the following areas are anticipated to improve with further research: To improve the projected accuracy of SoC and SoH, researchers are looking into more innovative artificial intelligence techniques and improving their training algorithms. The best search techniques for the optimal number of neurons in the hidden layer must be explored and implemented into the neural network method to improve the neural network method's estimating performance.

Acknowledgments This publication is supported by award NPRP12S-0125-190013 from the QNRF (Qatar National Research Fund), a member of the Qatar Foundation. The information and views set out in this publication are those of the authors and do not necessarily reflect the official opinion of the QNRF.

References

1. Ranieri, M., Alberto, D., Piret, H., & Cattin, V. (2017). Electronic module for the thermal monitoring of a Li-ion battery cell through the electrochemical impedance estimation. *Microelectronics Reliability, 79*, 410–415. https://doi.org/10.1016/J.MICROREL.2017.06.010
2. Zhou, X., Stein, J. L., & Ersal, T. (2017). Battery state of health monitoring by estimation of the number of cyclable Li-ions. *Control Engineering Practice, 66*, 51–63. https://doi.org/10.1016/J.CONENGPRAC.2017.05.009
3. Brunner, D., Prasad, A. K., Advani, S. G., & Peticolas, B. W. (2010). A robust cell voltage monitoring system for analysis and diagnosis of fuel cell or battery systems. *Journal of Power Sources, 195*(24), 8006–8012. https://doi.org/10.1016/J.JPOWSOUR.2010.06.054
4. Wang, S.-L., et al. (2018). Open circuit voltage and state of charge relationship functional optimization for the working state monitoring of the aerial lithium-ion battery pack. *Journal of Cleaner Production*. https://doi.org/10.1016/J.JCLEPRO.2018.07.030
5. Wang, Z., Liu, K., Liu, J., Luo, Q., & Ma, C. (2017). Influence of the charging and discharging of the 18650 lithium-ion battery thermal runaway. *Journal of Loss Prevention in the Process Industries*. https://doi.org/10.1016/J.JLP.2017.11.008
6. Lu, Z., Yu, X., Zhang, L., Meng, X., Wei, L., & Jin, L. (2017). Experimental investigation on the charge-discharge performance of the commercial lithium-ion batteries. *Energy Procedia, 143*, 21–26. https://doi.org/10.1016/J.EGYPRO.2017.12.642
7. Yu, Q., Xiong, R., Lin, C., Shen, W., & Deng, J. (2017). Lithium-ion battery parameters and state-of-charge joint estimation based on H-infinity and unscented Kalman filters. *IEEE Transactions on Vehicular Technology, 66*(10), 8693–8701. https://doi.org/10.1109/TVT.2017.2709326
8. Xiong, R., Sun, F., Gong, X., & He, H. (2013). Adaptive state of charge estimator for lithium-ion cells series battery pack in electric vehicles. *Journal of Power Sources, 242*, 699–713. https://doi.org/10.1016/J.JPOWSOUR.2013.05.071

9. Xiong, R., Sun, F., He, H., & Nguyen, T. D. (2013). A data-driven adaptive state of charge and power capability joint estimator of lithium-ion polymer battery used in electric vehicles. *Energy, 63*, 295–308. https://doi.org/10.1016/J.ENERGY.2013.10.027
10. Lu, L., Han, X., Li, J., Hua, J., & Ouyang, M. (2013). A review on the key issues for lithium-ion battery management in electric vehicles. *Journal of Power Sources, 226*, 272–288. https://doi.org/10.1016/j.jpowsour.2012.10.060
11. Hoque, M. M., Hannan, M. A., Mohamed, A., & Ayob, A. (2017). Battery charge equalization controller in electric vehicle applications: A review. *Renewable and Sustainable Energy Reviews, 75*, 1363–1385. https://doi.org/10.1016/J.RSER.2016.11.126
12. Hannan, M. A., Hoque, M. M., Ker, P. J., Begum, R. A., & Mohamed, A. (2017). Charge equalization controller algorithm for series-connected lithium-ion battery storage systems: Modeling and applications. *Energies, 10*(9), 1–20. https://doi.org/10.3390/en10091390
13. Ju, F., Deng, W., & Li, J. (2016). Performance evaluation of modularized global equalization system for lithium-ion battery packs. *IEEE Transactions on Automation Science and Engineering, 13*(2), 986–996. https://doi.org/10.1109/TASE.2015.2434052
14. Khan, M., Swierczynski, M., & Kær, S. (2017). Towards an ultimate battery thermal management system: A review. *Batteries, 3*(1), 9. https://doi.org/10.3390/batteries3010009
15. Liu, K., Li, K., Peng, Q., & Zhang, C. (2018). A brief review on key technologies in the battery management system of electric vehicles. *Frontiers of Mechanical Engineering*. https://doi.org/10.1007/s11465-018-0516-8
16. Wu, C., Zhu, C., & Ge, Y. (2017). A new fault diagnosis and prognosis technology for high-power lithium-ion battery. *IEEE Transactions on Plasma Science, 45*(7), 1533–1538. https://doi.org/10.1109/TPS.2017.2706088
17. Lyu, D., Ren, B., & Li, S. (2019). Failure modes and mechanisms for rechargeable lithium-based batteries: A state-of-the-art review. *Acta Mechanica, 230*(3), 701–727. https://doi.org/10.1007/s00707-018-2327-8
18. Hu, X., Zhang, K., Liu, K., Lin, X., Dey, S., & Onori, S. (2020). Advanced fault diagnosis for lithium-ion battery systems: A review of fault mechanisms, fault features, and diagnosis procedures. *IEEE Industrial Electronics Magazine, 14*(3), 65–91. https://doi.org/10.1109/MIE.2020.2964814
19. Gandoman, F. H., et al. (2019). Concept of reliability and safety assessment of lithium-ion batteries in electric vehicles: Basics, progress, and challenges. *Applied Energy, 251*, 113343. https://doi.org/10.1016/J.APENERGY.2019.113343
20. Gandoman, F. H., Ahmed, E. M., Ali, Z. M., Berecibar, M., Zobaa, A. F., & Abdel Aleem, S. H. E. (2021). Reliability evaluation of lithium-ion batteries for E-mobility applications from practical and technical perspectives: A case study. *Sustainability, 13*(21), 11688. https://doi.org/10.3390/su132111688
21. Gandoman, F. H., van Mierlo, J., Ahmadi, A., Abdel Aleem, S. H. E., & Chauhan, K. (2019). Safety and reliability evaluation for electric vehicles in modern power system networks. In *Distributed energy resources in microgrids: Integration, challenges and optimization* (pp. 389–404). https://doi.org/10.1016/B978-0-12-817774-7.00015-6
22. Cheng, K. W. E., Divakar, B. P., Wu, H., Ding, K., & Ho, H. F. (2011). Battery-management system (BMS) and SOC development for electrical vehicles. *IEEE Transactions on Vehicular Technology, 60*(1), 76–88. https://doi.org/10.1109/TVT.2010.2089647
23. Xiong, R., Li, L., & Tian, J. (2018). Towards a smarter battery management system: A critical review on battery state of health monitoring methods. *Journal of Power Sources, 405*, 18–29. https://doi.org/10.1016/j.jpowsour.2018.10.019
24. Li, Y., Chen, J., & Lan, F. (2020). Enhanced online model identification and state of charge estimation for lithium-ion battery under noise corrupted measurements by bias compensation recursive least squares. *Journal of Power Sources, 456*, 227984. https://doi.org/10.1016/j.jpowsour.2020.227984
25. Li, X., Huang, Z., Tian, J., & Tian, Y. (2021). State-of-charge estimation tolerant of battery aging based on a physics-based model and an adaptive cubature Kalman filter. *Energy, 220*, 119767. https://doi.org/10.1016/j.energy.2021.119767

26. Li, S., Li, Y., Zhao, D., & Zhang, C. (2020). Adaptive state of charge estimation for lithium-ion batteries based on implementable fractional-order technology. *Journal of Energy Storage, 32*, 101838. https://doi.org/10.1016/j.est.2020.101838
27. Meng, H., & Li, Y.-F. (2019). A review on prognostics and health management (PHM) methods of lithium-ion batteries. *Renewable and Sustainable Energy Reviews, 116*, 109405. https://doi.org/10.1016/j.rser.2019.109405
28. Zhu, Q., Xu, M., Liu, W., & Zheng, M. (2019). A state of charge estimation method for lithium-ion batteries based on fractional order adaptive extended Kalman filter. *Energy, 187*, 115880. https://doi.org/10.1016/j.energy.2019.115880
29. Omariba, Z. B., Zhang, L., Kang, H., & Sun, D. (2020). Parameter identification and state estimation of lithium-ion batteries for electric vehicles with vibration and temperature dynamics. *World Electric Vehicle Journal, 11*(3). https://doi.org/10.3390/WEVJ11030050
30. Shen, P., Ouyang, M., Lu, L., Li, J., & Feng, X. (2018). The co-estimation of state of charge, state of health, and state of function for lithium-ion batteries in electric vehicles. *IEEE Transactions on Vehicular Technology, 67*(1), 92–103. https://doi.org/10.1109/TVT.2017.2751613
31. Chaoui, H., & Mandalapu, S. (2017). Comparative study of online open circuit voltage estimation techniques for state of charge estimation of lithium-ion batteries. *Batteries, 3*(2). https://doi.org/10.3390/batteries3020012
32. Zou, Y., Li, S. E., Shao, B., & Wang, B. (2016). State-space model with non-integer order derivatives for lithium-ion battery. *Applied Energy, 161*, 330–336. https://doi.org/10.1016/j.apenergy.2015.10.025
33. Tong, S., Lacap, J. H., & Park, J. W. (2016). Battery state of charge estimation using a load-classifying neural network. *Journal of Energy Storage, 7*, 236–243. https://doi.org/10.1016/j.est.2016.07.002

Chapter 3
Recognition of Electric Vehicles Charging Patterns with Machine Learning Techniques

Mohammadreza Shekari, Hamidreza Arasteh, and Vahid Vahidinasab

3.1 Introduction

In recent years, to supply some parts of the required energy, reduce fossil fuel usage, and decrease CO2 emission, the utilization of renewable energy sources (RESs) is highly attracted [1]. The presence of distributed energy resources (DERs), including RESs and electric vehicles (EVs), has changed the planning and operation of power systems dramatically [2]. Traditional power systems have been transformed into modern ones by increasing the penetration of DERs. Modern smart grids are becoming more reliant on smaller and decentralized generations than the conventional power systems powered by a limited number of large and centralized generation units.

The most popular categorization of EVs included hybrid EVs, plug-in EVs, and plug-in hybrid EVs. They are considered as the essential components of future smart grids because of environmental concerns, which led to the gradual disposal of conventional gas and diesel vehicles [3, 4]. In addition to environmental issues,

M. Shekari
Power Systems Group, Department of Electrical & Computer Engineering, Tarbiat Modares University, Tehran, Iran
e-mail: m_shekari@modares.ac.ir

H. Arasteh (✉)
Power Systems Operation and Planning Research Department, Niroo Research Institute (NRI), Tehran, Iran
e-mail: harasteh@nri.ac.ir

V. Vahidinasab
Department of Engineering, School of Science and Technology, Nottingham Trent University, Nottingham, UK
e-mail: vahid.vahidinasab@ntu.ac.uk

V. Vahidinasab, B. Mohammadi-Ivatloo (eds.), *Electric Vehicle Integration via Smart Charging*, Green Energy and Technology,
https://doi.org/10.1007/978-3-031-05909-4_3

due to the rising price of fossil fuels, EVs may also provide economic benefits to the users. Although the use of EVs has many advantages, increasing the number of these vehicles will be resulted in the energy demand increments and has become a new challenge for the grid operators to handle the electricity grid's balance.

Preserving the modern power grid's stability and resiliency requires proper coordination of RESs and EVs [5]. An uncontrolled EV charging pattern might wreak havoc on power distribution networks. There have been various studies on EVs' integration into smart grids. Generally, the researches around EVs and their impacts on energy systems are divided into three main domains:

- EV charging behaviors.
- Allocation of the EV charging stations.
- Coordination of the integration of the EVs with RESs.

The scope of this chapter is to use the machine learning (ML) techniques to investigate and analyze EVs' charging behavior. Firstly, an unsupervised learning method, i.e., K-means, is utilized in order to cluster the charging patterns of the EVs. The primary contribution of this chapter is to consider items that have not been well addressed in prior work. The most popular items for investigating the EV charging behaviors are start time, stay duration, and energy demand. Other items which might have significant effects on the classification of the EVs are traffic patterns and weather that will be addressed in this chapter. Finally, a supervised learning method, i.e., K-NN, is utilized for further classification of the dataset of the EVs.

This chapter aims to provide readers with a better knowledge of EV charging habits. Hence, researchers who want to improve the planning of smart charging infrastructure and exploitation of smart charging technologies can benefit from the results of this research. Figure 3.1 illustrated a graphical abstract that highlights various parts of this chapter.

3.1.1 Electric Vehicles

Environmental issues, resource depletion, and energy dependence have given rise to changing the structure of transportation and replacing EVs [6]. Various estimations have been considered for the penetration level of EVs. As an instance, there are approximately 32 million EVs currently in the UK, and it's predicted that more than 112 million of them will be in 2050 [7].

Currently, there are different types of EVs around the world, considering the manufacturing technology. But, generally, they can be divided into five groups: battery electric vehicles (BEVs), plug-in hybrid electric vehicles (PHEVs) [8], hybrid electric vehicles (HEVs), fuel cell electric vehicles (FCEVs) [9], and extended-range EVs (ER-EVs) [10].

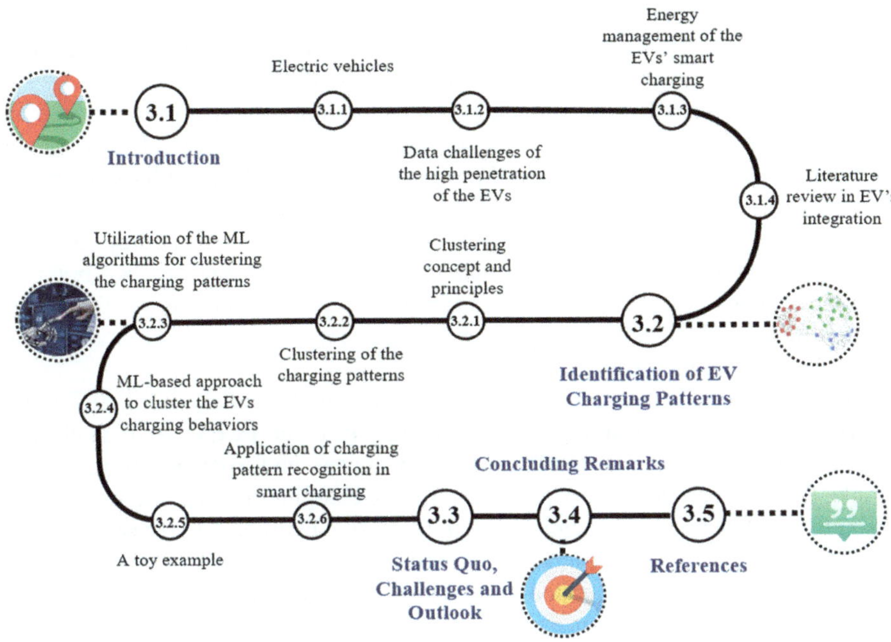

Fig. 3.1 Graphical abstract

3.1.1.1 Taxonomy of EVs

BEVs are powered entirely by electricity. Therefore, BEVs are referred to as full electric vehicles. BEVs neither have an internal combustion engine nor use any liquid fuel. Instead, BEVs usually use large battery packs to provide the vehicle with enough propel. BEVs have much larger batteries and kilowatt-hour (kWh) outputs than comparable HEVs and PHEVs, because they rely solely on electricity. In addition, BEVs often cost more than other types of EVs due to different battery technologies. BEVs require a charge to be driven. This can be accomplished via a home charger or a fast-charging station or by recovering energy through regenerative braking [11].

Moreover, a conventional combustible engine, in collaboration with an electric engine powered by a pluggable external electric source, propels PHEVs. They are comparable to HEVs in this regard. PHEVs often contain larger battery packs and more powerful electric motors than HEVs, because their electric system conducts a lot of heavy lifting while driving. This means that PHEVs can be moved fully on electricity with the internal combustion engine turned off. Driving a PHEV is comparable to driving an HEV because the vehicle will automatically recharge the battery and, depending on the conditions, switch between internal combustion and electric power. In contrast, PHEVs may be filled up with both fossil fuel and electricity. A PHEV can run solely on gasoline if all of the battery charge is used up.

Vice versa, it can run exclusively on the battery charge if all of the fuel is used up [12]. As a result, PHEVs could be able to store sufficient power from the network to cut gasoline usage considerably under normal driving circumstances [13].

HEVs are propelled by a combination of an internal combustion engine and an electric motor which also reduces fuel consumption. Unlike PHEVs, HEVs are not grid-connected. Instead, the produced power from the vehicle's combustion engine and brakes are used to charge the electric motor's battery. HEVs are the most likely classic internal combustion engine vehicles to drive because they can only be refueled with traditional fuels (usually petrol). Through regenerative braking, HEV technology charges the battery automatically [14]. When the conditions are right, it turns on the electric motor system, so drivers don't have to keep an eye on the charge or plug their cars into outlets.

On the flip side, FCEVs are similar to BEVs in that they solely utilize electricity to drive, but their energy storage is considerably different. FCEVs have an electric engine that runs on a mixture of compressed oxygen and hydrogen from the air. Unlike BEVs which store electrical energy from a charging station, FCEVs generate their electrical charge via a chemical reaction utilizing hydrogen. As a result, FCEVs can now be filled with hydrogen and do not need to be charged from the grid [15]. Because the only waste generated by this procedure is water, it can be considered environmentally friendly. However, most of the hydrogen used in this EV type is extracted from natural gas [10].

ER-EVs are very similar to the BEV category in terms of design. However, the ER-EVs also come with a backup combustion engine that can charge the vehicle's batteries if necessary. Unlike those embedded in PHEVs and HEVs, this type of engine is solely utilized for charging, and for this reason, it is not connected to the vehicle's wheels. An ER-EV has an auxiliary power unit (also known as a range extender) that extends the vehicle's range. The majority of range extenders are small internal combustion engines that power an electric generator, which provides power to the electric batteries and motor. When the ER-EV's small range extender motor runs, CO2 is produced, but not when the ER-EV runs on electric power. As a result, an ER-EV will produce substantially less CO2 over its lifetime than a vehicle powered by an internal combustion engine [16].

These types of EVs are illustrated in Fig. 3.2.

3.1.1.2 EV Integration's Benefits

What all types of EVs have in common is that they are charged with electricity, and it is possible to use the stored energy for driving up to a certain distance. EVs' integration with the power grid has various benefits. It is believed that the large penetration of EVs may assist in minimizing greenhouse gas emission, and they can be regarded as the pioneers in offering a pure form of transportation [6]. Also, EVs can enhance the integration of RESs. Increased size (or capacity) of the energy storage system (ESS) is required for the large-scale integration of RES [8]. As a

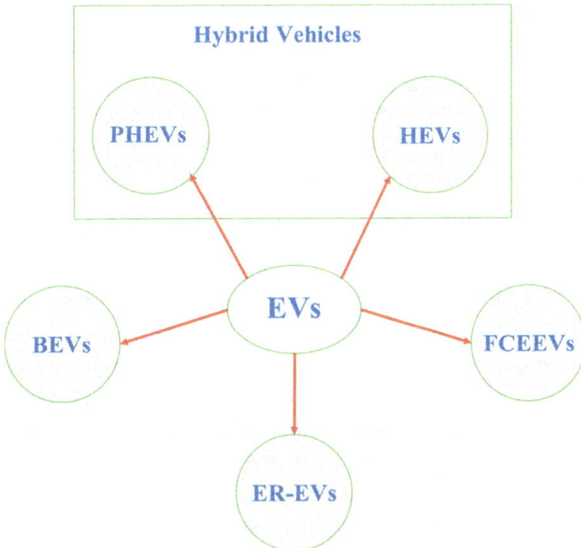

Fig. 3.2 Five types of EVs

result, EVs may also be used as storage devices to overcome the problems of RESs [10].

What made EVs special is their ability to exchange electricity with the power grid in vehicle-to-X operation modes, such as unidirectional control of vehicles (V1G), vehicle-to-grid (V2G), grid-to-vehicle (G2V), vehicle-to-home (V2H), and vehicle-to-building (V2B). As we are currently heading toward EVs to reduce greenhouse gas emissions, these functions of EVs may also operate as a dynamic-natured ESS, delivering stored energy in their batteries back to the power network. Besides, because EVs spend so much time parked [17], they may be utilized as ESSs without causing drivers any difficulty.

Furthermore, EVs can be utilized to establish the balance between generation and consumption while preserving the stability and quality of the grid [18]. Also, EVs can offer several possible ancillary services to the grid, including frequency regulation, load leveling, spinning reserve, voltage regulation, and congestion reduction at transmission and distribution feeders [9, 19]. EVs can also improve energy efficiency [8] and engage in energy trading, providing an income stream for drivers and aggregators to offset the battery degradation caused by V2G participation [20].

3.1.1.3 Challenges and Problems of EVs High Penetration

Despite the aforementioned benefits from EVs, the high penetration of EVs also poses some problems for power system operators due to the high-power needs and caused negative impacts on the power grid. The electrical network will be strained as a result of the load generated by EV charging at different charging stations.

This could bring negative impacts such as voltage deviation, transformer saturation, power loss, voltage regulation problems, and voltage fluctuation [21–23]. When the number of EVs grows, so does the overall load demand, while the voltage constantly decreases. Therefore, in the case of the high penetration of EVs, voltage boundaries are violated, affecting the power network security. Also, due to the existence of power electronic equipment, charging an EV can cause power quality difficulties, such as harmonics and voltage imbalance [24, 25]. For instance, voltage profile problems are more likely to occur in radial networks, especially if peak demand periods overlap with EV charging periods [8]. To address this problem, a high-quality filter is used to connect the charger to the power grid, preventing harmonics from affecting the grid as well as a motor connected to the vehicle [26].

Furthermore, uncoordinated EV charging behavior can result in extra deterioration and instability in power distribution networks [6]. When EVs are charged in large numbers simultaneously (dumb charging, uncoordinated charging), the demand for electricity rises, producing an imbalance in the power grid [27]. Besides, it is difficult to expand charging station capacity to meet the increasingly changing demands due to the implications of the power constraints.

Moreover, EVs' high penetration also causes challenges with the quality of distribution network, such as off-nominal frequency problems, network congestion, and three-phase voltage imbalance. Because EVs are single-phase mobile loads, they may be plugged into at any of the three phases of distribution networks at any time. It may result in a situation that one phase's electrical components, such as overhead line, power supply cable, or transformer, are severely loaded, while the other phases are not. Unbalanced three-phase loads can result in various power quality problems, including transformer failures, equipment failure, and relay malfunction [19].

Two of the large-scale EV integration major problems are grid overloading and load forecasting. Overloading the grid may lead to voltage control problems, voltage fluctuations, increased peak demand, decreased reliability and efficiency, and increased temperature of the lines. Large-scale EV integration also has a significant impact on load forecasting. All of these difficulties impact the system's overall efficiency, which is incompatible with the development of EVs and EVs' charging stations [23]. In the electrical distribution system, load forecasting is critical for estimating how much electricity is generated by calculating peak demand and baseload. However, since introducing EVs and charging stations, the complexity of load forecasting has been increased because estimating the changing loads was a challenging problem [28]. Researchers suggested various solutions to overcome these types of problems. For instance, state of charge (SoC) levels and battery management system (BMS) can be utilized using the Internet of things (IoT), which could result in more accurate load forecasting [23].

EVs' charging time and traffic congestion management at charging stations are also among the EVs' high penetration problems. The majority of EVs require a long time to charge, which is inconvenient [29]. Because of the pressure on the grid and physical space restrictions, the straightforward approach of expanding charging stations to enhance charging capacity for overcoming the issue of longer charging time is ineffective [30]. As a result, scientists have concentrated on creating smart scheduling algorithms that employ modeling and optimization to control the demand for public charging [31, 32]. Energy management of EVs' charging significantly influences the wholesale power market, emphasizing the need to know charging behavior [33].

Another issue that has constantly challenged power system operators is the uncertainty in RESs and EVs. Charging operations for EVs are carried out in a highly distributed and dynamic context, with uncertainties arising from charging time, charging station availability, EV arrivals, and energy costs. Moreover, because EVs are very spatially and temporally unpredictable, it is challenging to manage EVs as new loads while preserving grid reliability and security [22, 30, 34, 35]. Based on the optimization procedure's objective, these uncertainties will impact decisions about when, where, and how much to charge an EV. Furthermore, there are additional uncertainties in power systems, including the electrical grid's status, the generation of RESs, network congestion, availability of charging station, and the quantity of EVs accessible to deliver V2G services [21].

A summary of the mentioned challenges and problems due to the high penetration of EVs is given in Table 3.1.

So far, data challenges have not been addressed, which is one of the most important challenges of the high penetration of EVs. Due to the high importance of this matter, data challenges caused by the large-scale usage of EVs, as well as the solutions to deal with these problems, will be presented in the following.

3.1.2 Data Challenges of the High Penetration of the EVs

The big data problem is defined by the massive volume of data created by devices, EVs, buildings, the power grid, and many other connected objects and increasing data transmission speeds [36]. Smart grid and EVs, among many other sectors linked with the IoT, face this problem since they are both producers and consumers (i.e., prosumers) of extensive data. Moreover, travel records and information obtained through sensors embedded in vehicles are among the data generated by EVs. Users' driving behaviors, battery charge rates, acceleration, location, tire pressure, battery security data via a BMS, and grid charge management data via charging stations are all examples of data that EVs create and store continually. Smart gadgets and wearables are also carried by drivers, which add to the data collected on the road [37].

Also, some parts of EV data are from a variety of sources, including batteries and onboard chargers. Onboard electronic control units and BMS provide the majority of EV data. Most charging/discharging decisions are based on the SoC of EV batteries. SoC data and how an EV battery is functioning are displayed in BMS logs [36]. In addition to the data directly collected from EVs, drivers can also voluntarily contribute to collecting driving patterns and charging habits information.

Table 3.1 EVs high penetration challenges

Challenges and problems	Effects	References
The strain on the electrical network	Voltage deviation, transformer saturation, power loss, voltage regulation problems, and voltage fluctuation	[21–23]
Violation of voltage boundaries	Negatively affecting the power network security	[24, 25]
Voltage profile problems in radial networks	The affected grid, as well as the motor connected to the EV, causes harmonics	[8, 26]
Uncoordinated charging/dumb (non-smart) charging	Extra deterioration and instability in power distribution networks	[6, 27]
The strain on the power electronic equipment	Power quality difficulties such as harmonics and voltage imbalance	[24, 25]
Distribution network quality, such as network congestion, off-nominal frequency problems, and three-phase voltage imbalance	One-phase electrical components, such as overhead line, power supply cable, or transformer, are severely loaded, while other phases are not. Unbalanced three-phase loads can result in various power quality problems, including transformer failures, equipment failure, and relay malfunction	[19]
Grid overloading	Voltage control problems, increased peak demand, decreased reliability and efficiency, and increased temperature of the lines	[23]
Load forecasting	Load forecasting has grown more complex because calculating changing loads was a challenging problem	[23, 28]
Prolonged EVs' charging time and difficulties in traffic congestion management at charging stations	Because of the pressure on the grid and physical space restrictions, the straightforward approach of expanding charging stations to enhance charging capacity for overcoming the issue of longer charging time is ineffective. This problem also has some negative effects on the wholesale power market	[29]–[33]
Uncertainties, including the electrical grid's status, the generation of RESs, availability of charging station, EV arrivals, energy costs, network congestion, and the quantity of EVs accessible to deliver V2G services	These uncertainties will impact decisions about when, where, and how much to charge an EV. Also, it is challenging to manage EVs as new loads while preserving grid reliability and security	[21, 22, 30, 34, 35]

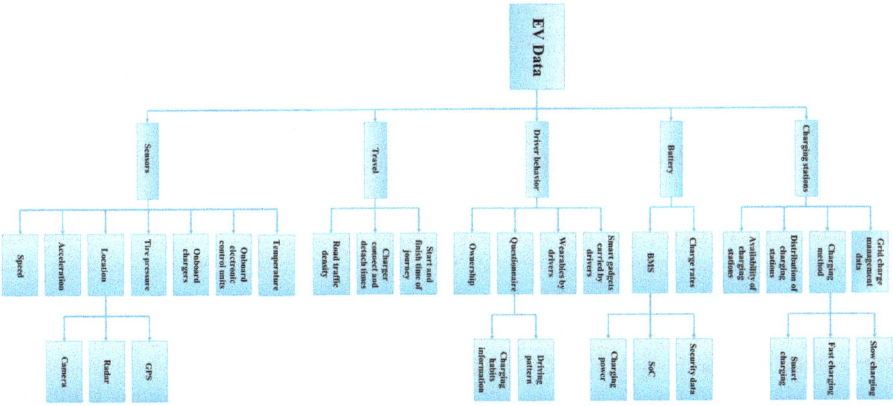

Fig. 3.3 EV data

Furthermore, trip information, including start and end times of journeys, charger connect and detach times, and battery SoC may be readily gathered.

Various types of data have been employed, such as road traffic density, distribution of charging stations, and EV ownership. However, as cities get smarter, such data will grow in bulk, and mining them alongside EV data will provide further challenges for smart grids. Generally, EV data can be classified as shown in Fig. 3.3.

However, there is no consensus on how much data a connected EV might create daily; some estimates put the maximum limit at 32 terabytes [38]. Millions of EVs expected to be on the road over the next decade will lead to the generation of a massive amount of data, which is challenging to be handled. Because the data acquired from EVs is diverse and large in volume, standard statistical methods for building a model (to study and analyze the EV problems) may not be effective.

Data science should be widely employed today to address a variety of EV-related issues. EVs' data is produced from multiple mentioned sources, including EV charging stations, vehicles, and road sensors. Moreover, the industry is increasingly providing private data to researchers, paving the way for developing high-quality data-driven research. Therefore, on the one hand, data-driven approaches to EV analysis are required. On the other hand, various new research issues cannot be solved using existing techniques. Management of smart charging stations, which includes expanding, suspending, and reallocating charging stations, is an example of these issues [39].

In this section, several EV-related big data challenges were mentioned. Once these challenges are overcome using data analysis techniques and data-driven methods, there is a chance to go toward more intelligent approaches such as smart charging to handle large-scale integration of EVs. Addressing energy efficiency issues, developing policies for allocating charging stations, assessing the capability of power distribution networks to manage increased charging loads, and estimating the market value for the services offered by EVs (i.e., V2G opportunities) are among other advantages of this phenomenon [36].

3.1.3 Energy Management of the EVs' Smart Charging

It's important to figure out when the EV should be charged and when it should be utilized as energy storage during peak load times to offload the energy grid. It's also crucial to managing the EV battery to meet the demands of both the driver and the energy grid [8, 39]. There are two main strategies in this regard: (1) dumb (non-smart) charging and (2) smart charging [8, 40, 41].

EVs start charging as soon as they are connected to the charging station in the dumb charging scenario. The time it takes to connect does not have a fixed value; instead, it is determined by the normal distribution of return following transportations [8].

The smart charging scenario is about automatically managed charging and necessitates installing intelligent vehicle controllers that also control the EVs' charging. This is set in conjunction with smart controllers placed on the load and production units. Charging pursues the load demand curve, and in the following, EVs are charged during off-peak hours. Thus, EV charging rates increase as grid demand decreases. To put it another way, this is a case concerning valley filling.

In other words, smart charging refers to adjusting the charging cycle of EVs to the power system's circumstances, as well as the demands of EV drivers. This makes it easier to integrate EVs while fulfilling mobility requirements. As a result, smart charging is a method for optimizing the charging process depending on the local RES availability, the constraints of the distribution grid, and users' priorities [42].

Smart charging stations should determine the quantity of energy provided to each plug-in EV during each period, as restricted by the power distribution networks and energy storage systems. Thus, energy management is essential for coordinating EV transit and charging at central smart charging points. In order to deliver high-quality services and maintain the power grid's stability simultaneously, energy consumption for each journey, the timing of each charge, and the capacity of each charging station should be addressed [22]. Furthermore, smart charging facilities should be able to cooperate with the intermittent behavior of the RESs, including solar panels or wind turbines, which might cause uncertainties and problems to smart charging scheduling.

3.1.3.1 Concepts and Applications

Smart charging is commonly considered in the smart grid environment and offers a degree of control over the process of charging [43]. Smart charging generates charging schedules for each EV, with the goal of effectively and equitably allocating charging capabilities across a number of EVs. However, practically EV charging methods use charging profiles like constant voltage and/or current [44]. Therefore, to prevent gaps between charging plans and actual EV power usage, smart charging should take charging profiles into account. Besides, smart charging includes different pricing and technical charging options. Also, at a larger penetration level and for the delivery of ancillary services, as well as real-time balancing, more advanced smart charging techniques (i.e., direct control techniques) would be required as a long-term solution [45].

Moreover, charging point operators, EV drivers, and the other components of the EV market behave independently of the electricity market in the non-smart charging strategies. Against dumb charging, smart charging requires strong coordination between the electricity market and electric mobility to satisfy EVs' integration requirements in the power grid while maintaining its flexibility. To put it another way, smart charging is the process of charging an EV managed by two-way communications among multiple actors to optimize grid management, customer needs, and energy generations, such as RESs with concerning limitations, costs of the system, security, and reliability.

There are a variety of smart charging mechanisms. Some of the most important forms are as follows [19, 42]:

1. Unidirectional control of vehicles (V1G): Allows to increase or decrease the charging rate. In this mechanism, vehicles or charging infrastructure alters their charging speed.
2. Bidirectional vehicle-to-grid (V2G): Permits the EV to deliver services to the grid in the discharging mode. In fact, in this case, the smart grid manages EV charging while it is possible to return power to the grid.
3. Vehicle-to-building (V2B) and vehicle-to-home (V2H): They are additional types of two-way charging in which EVs are utilized as a residential backup power generation during electricity outages. In addition, in this method, EVs can also be used to increase the self-consumption of energy produced on-site. In other words, vehicles will serve as backup power sources for homes or buildings in this scenario.

However, among the various forms of smart charging, the V2G method is the most welcomed and used [19, 23, 41, 42, 46–48]. Smart charging enabled by the V2G method controls EV loads. It involves consumer reaction to pricing signals, as well as automatic responses to control signals reacting to network/market circumstances, or the combination of them. This could be done while taking into account the customer's desire for EV availability. However, it entails delaying certain charging cycles or regulating power in response to certain restrictions, such as connection capacity, user demands, and real-time local energy generation.

Furthermore, smart charging of EVs has many advantages and applications, such as network congestion management, minimizing grid infrastructure investments, ancillary service provision, and peak shaving. It also reduces the impact of EVs' load, allowing more solar and wind power to be used for providing flexibility.

Moreover, EVs could modify their charging patterns using smart charging to fill load valleys, reduce peak load, and assist the real-time grid's balancing by changing their charge levels. Using EVs as a flexible resource through smart charging techniques could minimize the demand for carbon-intensive fossil fuel power plants to balance renewables [49]. Besides, the prolonged charging time challenge can be solved using a smart charging approach in which the smart grid collects real-time data about loads at different charging stations and passes it on to each EV. Thus, it makes it possible for EVs to develop charging strategies (i.e., the optimum charging station along the route to the destination) [31, 50].

In addition, existing parking lots are being transformed into smart charging parking lots with Internet access for slot booking and traffic information at parking locations [51]. In fact, smart charging may be used to quickly pre-book charging station spaces and monitor the status of available slots. This method makes load forecasting more efficient and would ease the communication with the RESs, whether at home, at the workplace, in a parking lot, at a shopping center, or at a charging station [52]. Besides, distribution system operators (DSOs) may profit from the smart charging of EVs in a variety of ways. For instance, it can assist in managing capacity constraints via demand response or serve as a balancing resource to accommodate DERs within a distribution system [53]. Smart charging also decreases the costs of strengthening local power networks and diminishes the load peaks.

On top of those, reducing RES curtailment, enhancing RESs integration (in system operation and long-term expansion planning), improving local utilization of RESs, cutting peak loads, avoiding high investments due to the peak demands, and mitigating grid reinforcement needs are among other applications of EVs' smart charging.

A framework for designing a smart charging system for integrating the EV-RES system into the grid is illustrated in Fig. 3.4 [19].

3.1.3.2 Challenges and Opportunities

To allow the implementation of energy management of EVs' smart charging, intelligent sharing of information, standardized communication protocols, information of connected EVs, and charging stations utilizing smart meters and intelligent infrastructures are required. Some of the key factors to enable the deployment of smart charging are as follows [45]:

1. *Charging infrastructure:* Charging infrastructure development necessitates large expenditures. Since there are currently few economic models for private investment at the moment, governments should catalyze the installation of charging stations either in private or public areas. Furthermore, a key challenge is figuring out how to effectively charge, collect, and aggregate the EV load on the grid. This would influence critical choices in the developing charging infrastructures, such as the location of charging stations, which technology to employ, and how to mix slow smart chargers with fast smart chargers in order to satisfy consumers' urgent requirements.

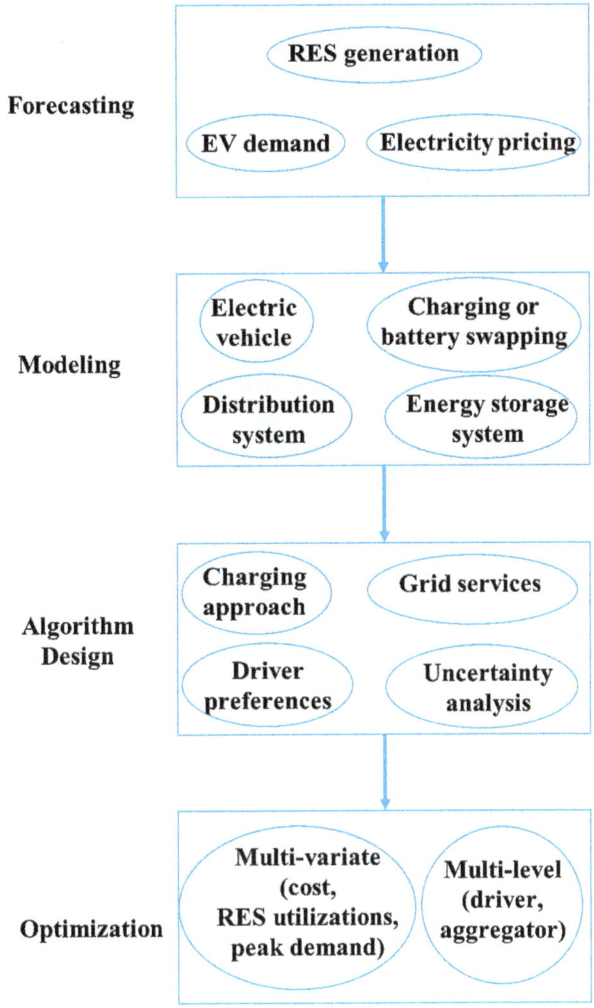

Fig. 3.4 A framework for a smart charging system [19]

2. *Definition of stakeholders' function and responsibilities:* A utility, energy supplier, e-mobility organization, or even a firm specializing in demand response services may regulate charging a fleet of EVs to unlock the potential of V2G use cases and smart charging plans.
3. *Regulation design of EV's integration with grid:* Smart charging and V2G technologies will not be fully implemented without the right incentives in the form of dynamic price signals and without the possibility to stack money from different revenue streams, respectively. This requires the existence of well-functioning energy markets. However, competitive wholesale markets, retail markets, and emerging EV markets are not always in place. Even if such markets

exist, their structure needs to be evolved, and regulations should be modified to enhance the motivation for the evaluation of services provided by EVs.
4. *Communication protocol:* Communication protocols should be established to optimize the system and allow information sharing across all participants.
5. *Artificial intelligence and big data methods:* Artificial intelligence (AI) and big data advancements could help smart charging systems provide better services. Smart charging capabilities would be enhanced by ICT developments, such as data management and analytics from drivers, charging habits, and charging stations. Besides, data analytics and digital technologies will make it feasible to fix mobile demand and power supply patterns as closely as possible and determine the best sites for charging stations.

The introduction of big data analytics and ML has had significant impacts on current scientific problems. For example, it has transformed natural language processing, picture, audio, and video recognition. In particular, it can be seen that the focus has been changed to use data-driven techniques to tackle the EV smart charging challenge. Algorithms of ML could be trained and learned trends and patterns through previous data of charging load and user behavior [6, 54]. ML-based scheduling, forecasting, and clustering strategies are widely deployed to handle high EV penetration and enhance smart charging deployment. These solutions are primarily designed to reduce the impact of EV charging on electricity distribution.

The goal of the EV charging schedule is to keep the electricity network stable by ensuring that the power generation and energy consumption of electricity are balanced. In addition, scheduling EVs' charging is a practical approach to shift the demand, reduce extra-generation problems, and diminish peak challenges [55].

When it comes to forecasting techniques, network stability and fault prevention are highly dependent on daily demand forecasts. As a result, the accuracy of the EV charging forecast is critical for utility and decision-makers. Moreover, implementing an exact forecasting model would support the development of EV charging, improve prediction precision for optimal dispatching, and motivate manufacturers to promote EV usage.

Furthermore, to identify the most frequent load patterns, clustering methods were applied to EV smart charging. In a multivariate dataset gathered from the field, the clustering of EV charging is employed to find clusters of identical charging items. Important patterns linked with distinct forms of demand over the course of a year, month, or week could be recognized by clustering these statistics across that interval [56].

3.1.4 Literature Review on EV Integration

There have been various studies on EVs' integration into smart grids. Generally, the researches around EVs and their impacts on energy systems could be divided into three main domains:

- EV charging behaviors.
- Allocation of the EV charging stations.
- Coordination of the integration of the EVs with RESs.

In this context, EV penetration in California was modeled using the regulated and optimized charging options [57]. Based on this study, the authors concluded that in smart charging mode, EV charging is managed that has resulted in just a 1.3% increase in peak demand, compared to an 11% increase in an uncontrolled charging manner. Similarly, researchers in [58] claimed that switching from uncontrolled charging mode to schedule charging mode could decrease the peak load increment from 19% to 0–6% when EVs integrate into the grid. They also suggested that charging can avoid increasing peak demand if it would happen only at off-peak hours.

Moreover, according to [59], if DSOs do not recognize shortages in their grids, the grid infrastructures may limit the high penetration of EVs in residential and private regions. Thus, they should integrate smart charging, taking into account not just time limits but also location constraints. Besides, the impact of local EV penetration on peak load of the German power grid in the presence and absence of smart charging is investigated in [60]. Results showed that with utilizing time-of-use tariffs and the V1G method, the peak demand could diminish by 16%.

A summary of the current existing standards for grid integration, EV charging, and safety is discussed in [61]. Besides, the EV charging infrastructures, such as the power, control, and communication infrastructure, are presented, and the effects of EV integration on different aspects of power systems are also examined. In addition, the authors of [62] investigated the EV integration with power systems based on their relevance to different players in the electricity market, including DSOs, generation companies, EV aggregators, and drivers using classification and scheduling methods [63]. have presented the distributed resource allocation (DRA) strategy for EV integration into the power grid based on the concept of output agreement. It also recommended an appropriate charging method for each grid-connected EV to meet grid objectives and EV drivers' economic and social interests.

Furthermore, V2G's idea, structure, benefits, difficulties, and optimization techniques are discussed in [64]. In addition, the authors examined the advantages, services, and potential hurdles of using V2G technology, as well as different V2G optimization approaches and new insights into V2G technology possibilities. Reference [65] investigated the interactions among charging, on-road power management, and battery-degradation mitigation of EVs to overcome the energy management challenges of EVs in a smart grid environment. Besides, [66] investigates how EVs might help smart cities to establish sustainable energy as a service business model. According to this research, EVs are valuable assets for a sustainable energy future since EV batteries provide the potential to store electricity generated from RESs.

In addition to such researches, many practical projects for integrating EVs with the power grid are being carried out worldwide. For instance, San Diego Gas & Electric (SDG&E) has started an EV-grid integration pilot project that aims to improve grid stability by making fleets of EVs available as dispatchable DERs [67]. The company used an app to match customers' preferences with dynamic pricing.

Nuvve[1] is another V2G technology pioneer. Since 2009, Nuvve has provided a wide range of power system services, such as frequency and supply reserve capacity, in several energy markets in Pennsylvania, Jersey, and Maryland in the USA [68]. Customers just have to give information about their SoC status and whether or not they utilize EV. This is the program that determines whether power should be returned to the grid or another service should be provided. Besides, Nuvve recently has announced plans to deploy 1500 smart chargers with V2G capabilities in the UK. Moreover, this company, along with the other EV-grid integration pioneers including Enel,[2] Insero,[3] Nissan,[4] and Mitsubishi,[5] took part in the Danish parker project aiming to use smart charging technology and depend on the collaboration among the power and automotive industries to show the EVs' capability for supporting power systems based on the RESs [69]. The project demonstrates that V2G can play a major role in boosting vehicle income and providing grid flexibility. However, there are still some practical issues to be overcome, such as uncertainty regarding battery deterioration, communication standards, and consumer awareness of the V2G system.

3.2 Identification of EV Charging Patterns

To build a powerful model for analyzing the utilization of EV smart chargers, it's essential to study EV owners' charging behavior to validate the underlying dataset and understand regularities and patterns in the dataset [70]. EVs may adopt different charging patterns than others in order to overcome obstacles in using the charging infrastructure, such as long charging periods. Therefore, understanding what elements contribute to this charging behavior can aid in the development of methods to promote a more efficient charging system [71].

Moreover, the preferred charging patterns of customers should be considered when predicting the implications of new demand on the grid constraint. By examining consumers' preferred charging times, it is feasible to estimate how the extra electricity consumption will be distributed all over the day. Furthermore,

[1] Nuvve is a publicly traded company accelerating transportation electrification through its proprietary V2G technology and offering charging and grid services.

[2] Enel (Ente Nazionale per L'energia Elettrica) is an Italian multinational manufacturer and distributor of electricity and gas.

[3] Insero is a company that wants to create sustainable growth and development within the energy, information technology, and EVs in Denmark's local area.

[4] The Nissan Motor Company is a Japanese multinational automobile manufacturer headquartered in Nishi-ku, Yokohama, Japan.

[5] Mitsubishi Motors Corporation is a Japanese multinational automotive manufacturer headquartered in Minato, Tokyo, Japan.

by monitoring consumer preferences, it is possible to determine the electricity consumption on each EV charging station at a specific time of day. As a result, analyzing the charging patterns of EVs is critical for precisely calculating the additional grid constraint posed by EVs and establishing a roadmap for future energy regulations [72].

Besides, it is critical to understand consumers' different EV charging patterns to provide sufficient EVs infrastructures. In other words, consumers' electricity consumption patterns have an impact on peak electricity demand, which is a key factor in determining the amount of power capacity required. Exploring EV charging patterns is essential not just for policymakers and suppliers but also for EV manufacturers because these charging infrastructures are a surefire way to attract new EV purchasers [73].

EV charging patterns could be quantified based on daily driving patterns and individual activity schedules [74]. In other words, a charging pattern is a representation of a customer's preferred charging time, location, and EV charging station type. Customers' charging patterns contain information including the average daily trip, types of destinations, driving motives, preferred charging times of a day, and charging stations. Charging patterns analysis included recognizing electricity consumption by the time of a day and investigating the desire to utilize public and residential EV charging stations [73].

In addition, different charging patterns, including single charge, multi-charge, and deferrable or partial charge with configurable energy demands, will add uncertainty to EV charging schedules while also providing flexibility [22]. Many factors influence EV charging behaviors, including travel patterns, energy usage, and charging infrastructure facilities. The EV usage behaviors and seasonal fluctuations have a significant impact on the charging frequency [75].

Vehicles will have multiple options for recharging during the day, taking into account the mobility pattern of each EV [32]. The time and location of charging for EVs depend on the EV type, its SoC, geographical area, and the availability of the charging station. Individual EVs have predictable charging patterns. However, depending on the business models, charging patterns of shared and commercial cars (for instance, taxi and other vehicle fleets) may be less predictable. Also, electric bus charging patterns are dependent on the charging station location.

Furthermore, new functionalities such as remote maintenance and management of charging stations will be enabled through data analytics and data management of charging patterns [45]. The distribution network operator could utilize the EV charging pattern as a guideline when developing future networks. In addition, charging patterns could also be utilized in load aggregation, clustering, forecasting, and direct load control strategies [74].

Two methods of EV charging pattern recognition are charging load prediction and charging pattern identification. Predictive and descriptive analytics are the terms used in data mining to describe these two types of analytics. In addition, many studies focus on consumer classification and clustering in order to improve usage predictions and demand balancing. Data mining algorithms like K-means clustering, self-organizing maps, and neural networks are commonly used in this case [76].

3.2.1 Clustering Concept and Principles

Clustering is a technique for grouping unlabeled data into clusters based on similarities between observations. The precision of this procedure is determined by the data characteristics (e.g., similarities between instances in the clustering technique). In other words, the higher the similarity, the more accurate the statistical procedure [77].

3.2.1.1 Concept of the Clustering

Clustering is one of the most common unsupervised learning methods. One of its goals is to reduce supervised learning's preprocessing noise. Clustering is the division of a collection of observations into clusters, each comprising items similar to one another but dissimilar to those in other clusters. In fact, clustering is the accumulation of data into groups of similar patterns [78–80].

Clustering is a basic method to perform supervised learning such as linear regression for each divided model resulting from unsupervised learning methods such as K-means. The K-means algorithm has been frequently employed because of its easy algorithmic design, quick clustering speed, and good clustering result [81]. In addition, clustering algorithms are used for various applications, including the prediction by regression and improving the efficiency of recommender systems [77].

3.2.1.2 Principles of the Clustering

In data mining, clustering analysis is a critical analytical algorithm. By using static classification, clustering divides similar objects into different groups or subsets, ensuring that the member objects in the same cluster have identical attributes [81, 82]. Cluster analysis of data also enables the detection of patterns and the reduction of data dimensionality [79].

Moreover, clustering analysis is a representative unsupervised learning method that classifies n items based on various predictor values observed for each object into clusters with similar characteristics and evaluates the links between clusters by finding the cluster features [83].

3.2.2 Clustering of the Charging Patterns

Clustering – as an unsupervised strategy – has been widely used for assessing load profiles to identify EVs with similar consumption patterns [84, 85]. Clustering algorithms have been applied to EV charging in order to identify the most prevalent and recurrent load profiles. The EV charging clustering approach finds groups of similar charging items in a multivariate dataset obtained from the field. The load profiles can be examined based on EV charging data to acquire a better understanding of the EV charging performance and energy demand patterns as well as to clarify the potential energy flexibility. It can assist grid operators in managing demand for electricity. The EV charging load curves are analyzed using the clustering approach. Clustering the charging patterns could be referred to as an unsupervised technique of data mining strategy that groups linked daily load profiles into separate groups. Clustering algorithms could also be utilized to develop load profiles that categorize EV charging behaviors into groups based on their commonalities [56].

Clustering algorithms are frequently employed to model energy usage because of two main reasons. Firstly, clustering is frequently utilized to analyze data and increase the prediction accuracy of energy models. Secondly, stable clusters can be utilized to organize, target, and interpret observed subjects if they are reproducible with respect to nonessential changes. However, clustering approaches are well recognized to be extremely sensitive to the algorithms and variables used. This can lead to erroneous assessments of predictive accuracy and misinterpretation of clusters in policymaking [86].

The goal of EV clustering is to find discrete clusters that capture some aspect of EVs' uniqueness [84]. By dividing the dataset into subgroups, the clustering process uncovers previously discovered dataset patterns [87]. For example, the daily EV charging demand analysis reveals the most common load profiles using clustering algorithms to put related profiles into the same subgroups [88]. The required dataset collection for EV clustering is usually operated with temporal and spatial datasets or a combination of them. The resolution of the generated typical load profile is directly affected by the temporal resolution of the data points [56].

In the literature on energy consumption, clustering approaches have been used extensively. For example, clustering has been used by various researchers to extract similar groups from large datasets. Examples include looking for groups of energy consumers with the same load or generation profiles and buildings [86]. Important patterns related to distinct forms of consumption can be differentiated by clustering the datasets across the year, month, or week [56]. In the case of EV charging patterns, finding patterns allows a better understanding and visualization of fleet utilization. Furthermore, the prediction problem's computation will be simplified if a set of common profiles is discovered. Thus, rather than estimating individual requirements, the number of vehicles in each cluster should be predicted [79].

Figure 3.5 depicts clustering in the context of EV charging patterns. Only two input features are used to group the objects in this simple example. We can observe four distinct charging behavior clusters based on the entrance time to the station and weekdays.

There are numerous advantages to clustering EV charging patterns. For example, the clustering approach seeks to explain the consequences of EV adoption on power system quality. These solutions offer a quick and efficient solution to the problem of charging station allocation. They also reduce overall operating costs by assigning

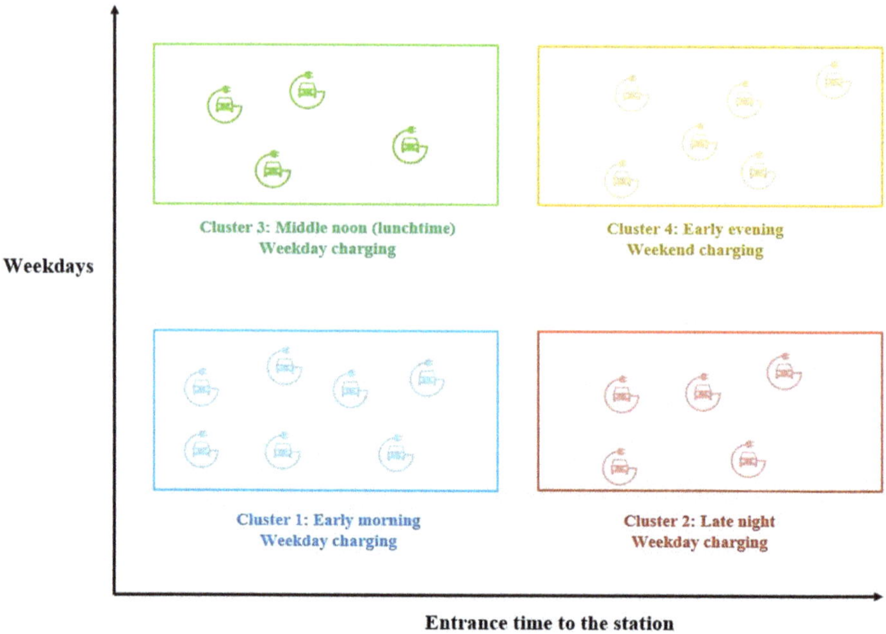

Fig. 3.5 Example of EV charging pattern clustering

the charging station to a selected charging demand cluster. However, some research gaps still exist regarding the clustering of EV charging patterns which may lead to potential risks for power system operators. For instance, the security and reliability aspects of clustering have not been addressed appropriately. In addition, according to the clustering research, various clusters can disclose different load profiles; therefore using the same clustering model to all of them may not be the best method [6].

3.2.3 Utilization of ML Algorithms for Clustering the Charging Patterns

Since the introduction of big data analytics and ML techniques, the focus has moved to use data-driven methods to overcome EV charging problems. An ML solution has three main advantages, including speed, generalizability, and scalability [89]. ML algorithms could be trained and learned trends and patterns using the historical data of the charging stations, as well as users' behavioral data.

AI and ML methods could be used to cluster and classify driver actions and forecast EV charging behavior [81]. ML allows computer systems to learn from their experiences without the requirement for explicit programming. ML algorithms utilize various datasets for training themselves. The models learn to recognize the dataset's trends and patterns over time. The models can produce accurate predictions and so provide predictive analytics after successful learning.

ML algorithms can be classified into the following categories based on the types of variables to be predicted (the response variable). On the one hand, if the response variable is continuous, the problem would be a regression problem. On the other hand, the problem is considered as a classification problem if the response variable is categorical [6]. ML algorithms are also divided into two main categories based on their applications: unsupervised learning and supervised learning.

3.2.3.1 Unsupervised Learning

The training dataset for unsupervised learning solely consists of inputs, with no labeled outputs. Labeling the data is considered a time-consuming and expensive process in many practical applications. The purpose of ML models is to discover structures or patterns in the data. Cluster analysis is an instance of unsupervised learning in which the ML model seeks out the groups of objects that share some common characteristics. To identify the clusters of EV behavioral patterns, unsupervised learning can be used.

For instance, K-means clustering, kernel density estimation (KDE), and Gaussian mixture model (GMM) are widely used in EV charging clustering as unsupervised learning methods. These methods are briefly introduced in the following:

- K-means clustering: Individual data points create k clusters in K-means clustering. First, each point is assigned to one of the k center points at random. Following that, using updated center calculations, the data points are transferred to the nearest center. It is necessary to know how many clusters there are; otherwise, a method should be employed for calculating it (such as the elbow approach as one of the best methods). The K-means clustering algorithm is simple, but it is sensitive to outliers and starting assignments [90]. Nevertheless, the K-means algorithm is one of the most widely used clustering techniques.
- Kernel density estimation: In parametric estimation approaches, the form of the probability density function (PDF) should be supposed. When this is not achievable, a continuous random variable's PDF could be estimated by kernel functions utilizing nonparametric estimation. Kernel functions should have a value of one for the region under the curve of the function. They also should be nonnegative and symmetrical. For example, diffusion-based KDE and Gaussian KDE are two popular kernels for KDE [91].
- Gaussian mixture model: The GMM could be considered as a probabilistic learning model which considers several normal distributions in the dataset to represent normally distributed subpopulations. Although GMMs are primarily employed for unsupervised learning, supervised learning versions are also available. Prior knowledge of the subpopulations is not necessary for an unsupervised model.

3.2.3.2 Supervised Learning

Labeled training datasets could train ML models in supervised learning. The input variables and the response variables are included in the dataset. The model repetitively learns the connection among the input and response variables using the optimization of a specified objective function.

A dataset comprising the vehicle's entrance time, name of the city, and vehicle's leaving time might be an example of supervised learning regarding EV charging patterns. The ML model would learn the association among the entrance time and name of the city as the inputs and the leaving time as the response variable, if the aim is to forecast the leaving time. The scope of this chapter does not contain a thorough examination of all supervised learning methods. However, some of the most commonly utilized algorithms for predicting EV behaviors are [6]:

- Support Vector Machine (SVM): SVM is primarily utilized to solve classification issues. However, it may also be utilized for regression problems. In that scenario, the technique is known as support vector regression. The optimum hyperplane that maximizes the margin between the various categories is used by the SVM to separate the classes. The inputs could be mapped to high-dimensional feature spaces[6] that could be linearly separable using kernels including polynomial, linear, and radial basis functions [92]. However, the lengthy training period is one of the SVM's major drawbacks. As a result, the SVM may be unsuitable for larger datasets.
- Decision Tree (DT): Both classification and regression problems can be solved with a DT. DTs use split points from the input features to break down difficult decisions into a series of more straightforward decisions. A decision node is a location where decisions are made. The leaf nodes are the sites where no additional splitting takes place. The average value of the entire components in the leaf node is used to predict regression problems. The set of projected classes is known as the leaf nodes in classification problems [93].
- K-Nearest Neighbor (K-NN): Although K-NN may be utilized for regression and classification, it is best known for classification. The K-NN does not require a separate training time. It is also known as a type of lazy learning method. A distance metric, commonly Euclidean distance, is utilized to discover the k closest neighbors of a new data point. Then it is given to the class with the greatest number of neighbors [94].

[6] High-dimensional spaces are used to model datasets with a large number of properties. A dataset can be represented directly in a space spanned by its features. Each record is represented as a point in space, and its location determined by the values of its attributes [95].

Fig. 3.6 Procedures for the clustering and classification of the EV charging patterns

3.2.4 ML-Based Approach to Cluster the EV Charging Behaviors

One possible approach for easing the stress of EV charging on the power grid is to predict EV charging behavior. EV consumers can be categorized into regular and irregular users. Regular users' charging patterns are predictable, whereas irregular users' charging patterns are unpredictable. As a result, it is helpful to model normal users' charging behavior, while irregular users' charging behavior will increase the prediction error.

Initially, the behavior of the EVs is clustered to examine the EVs' charging behavior. The following considerations are mainly considered while choosing variables as the basis for EV behavior clustering:

1. EVs have the most impact on the grid considering charging start time, charging end time, and charging demand.
2. The charging time is roughly proportional to the amount of charge.
3. Some users charge predictably, while others do not.
4. Weather affects road traffic congestion, and eventually, both of them influence the driving behavior of EV owners.
5. Weather will also affect the EVs' energy consumptions and thus their charging demands.

In this work, K-means clustering and K-NN algorithms are utilized to categorize the EV charging behavior. The procedures for the clustering and classification of the EV charging patterns are shown in Fig. 3.6.

3.2.4.1 Preprocessing

Data should be preprocessed before being clustered. The low-quality data in the raw data collection could be inspected and eliminated. For example, users with fewer than three records are deleted since they are unhelpful in determining behavior patterns. Weather data are also mapped to the proper time of charging events.

A linear normalization approach is used to normalize data, as illustrated in (3.1):

$$X^* = \frac{X - X_{\text{min}}}{X_{\text{max}} - X_{\text{min}}} \tag{3.1}$$

3.2.4.2 EV's Charging Behavior Clustering Using K-Means Algorithm

K-means is considered an unsupervised learning technique, meaning there are no labels on the data to be processed. It separates data points into k clusters, where each object belonging to the cluster corresponds to its nearest mean. The "means" is the representation of the mean of the data objects in the cluster, and "k" specifies the number of clusters. The algorithm assigns each point to the cluster corresponding to its nearest mean (the cluster center) and utilizes it as the clustering standard. This algorithm analyzes the historical charging data and generates assumptions of EV behavior for EV charging schedules.

The distance between vectors can be calculated using a variety of approaches. The most common method of measuring distance in the K-means algorithm is Euclidean distance, as formulated in (3.2):

$$\text{dict}\,(x_i, y_i) = \sqrt{\sum_{i=1}^{n} (x_i - y_i)^2} \tag{3.2}$$

where x_i and $y_i$are two points in an n-dimensional Euclidean space. Here, the mean and standard deviation of entrance and leaving times, the Pearson correlation coefficient between staying time and EV charging demand, and the Pearson correlation coefficient between weather attribute values and EV charging demand are used as the clustering criteria.

The initial cluster centers of the K-means method are randomly chosen from a pool of $n$data objects. The distances between the objects and cluster centers are then determined, and the associated objects are redivided according to the minimum distances. In the end, each cluster's mean is recalculated to become a new cluster center. The cluster center for the kth iteration is updated as follows:

$$\text{Center}_k = \frac{1}{|C_k|} \sum_{x_i \in C_k} x_i \tag{3.3}$$

where $|C_k|$ denotes the number of data points in the kth cluster and center k denotes a vector with D characteristics, as shown below:

$$\text{Center}_k = \left(\text{Center}_{k,1}, \ \ \text{Center}_{k,2}, \ \ \ldots, \ \ \text{Center}_{k,D}\right) \tag{3.4}$$

Furthermore, to redistribute the clusters and update the cluster centers, the K-means algorithm requires constant iteration. The group centroid locations are changed when all user group tags are updated based on the EVs who belong to, as demonstrated in the following equation:

$$\mu_j = \frac{\sum_{i=1}^{m} I\left\{c^i = j\right\} x^i}{\sum_{i=1}^{m} I\left\{c^i = j\right\}} \tag{3.5}$$

There are k centroids $\mu_j \in R^6, j \in [1, k]$. After clustering and before classification, the data results are normalized, and the user record is imported into the same tuple structure as the clustering centroids, as follows:

$$\mu = \left(\bar{t}_{\text{arrival}}, \bar{t}_{\text{departure}}, \sigma_{\text{arrival}}, \sigma_{\text{departure}}, \text{cor1}, \text{cor2}\right) \tag{3.6}$$

where cor1 and cor2 are the Pearson correlation coefficient among staying time and EV energy demand and the Pearson correlation coefficient between weather attribute values and EV charging demand, respectively.

The complete steps for K-means-based EV charging pattern clustering are shown in Algorithm 3.1.

Algorithm 3.1: EV charging behavior clustering using K-means algorithm

1: Generate K random numbers and corresponding K cluster centroids;

2: For each instance:

3: Classify each object with minimum distance;

4: Repeat

5: Update the centroid;

6: Reassign the instances to clusters;

7: Until no update.

However, in the K-means algorithm, determining the optimal number of clusters is always a challenge. There are several options for dealing with this issue. For instance, the elbow algorithm, which employs the sum of squared errors (SSE) approach, might be used to determine the most efficient number of clusters. The SSE rapidly decreases as the cluster number approaches the optimal number of clusters. However, SSE would continue to drop when the number of clusters exceeds the optimum cluster number. Meanwhile, the rate of the reduction of the SSE slows down. Therefore, the optimum K value can be better identified by plotting the curve of the K value on the SSE (K-SSE) and determining the inflection point on the way down.

3.2.4.3 K-NN Classification for EV Charging Behavior

There is a need to classify the behavior of new EV users after clustering the data using the K-means approach in order to manage and improve their charging behavior. However, re-clustering the entire dataset is inefficient if new data is entered. So, based on the results of the K-means clustering algorithm, the K-nearest neighbors (K-NN) algorithm could be applied to characterize the behavior of new EV user.

K-NN is a supervised learning approach for classifying new data by calculating the distance between new data points and previously marked data points. All of the selected neighbors are accurately classified using the K-NN method. It is a distance-based approach that determines the category of new data based on the category of its closest neighbors, such as when the distance between data points is calculated using the Euclidean method. Then, the new data is assigned to a group that has the least distance from that group's training data. However, the selection of k values is not governed by any specific rule. In general, a lower value is preferable based on the sample distribution, and an appropriate k value can be found via cross-validation.

The complete steps for K-NN-based EV charging patterns classification are shown in Algorithm 3.2.

Algorithm 3.2: EV charging behavior classification using K-NN algorithm

1: Calculating the distance between the labelled objects and the chosen objects;

2: Sorting the distance in ascending order;

3: Choose K objects with minimum distance from the selected objects;

4: Classify the objects that are repeated most often in the first $k$objects.

The ratio of the K-nearest training sample marker to the input marker can be used to calculate the training error rate (ER). The ER is calculated as (3.7):

$$\mathrm{ER} = \frac{1}{k \sum_{x_i \in N_k(x)} I\left(y_i \neq c_j\right)} = 1 - \frac{1}{k \sum_{x_i \in N_k(x)} I\left(y_i = c_j\right)} \tag{3.7}$$

Furthermore, the following coefficient in the training set could be maximized by selecting the proper K value:

$$\alpha_k = \mathrm{Max}\left\{\frac{1}{k \sum_{x_i \in N_k(x)} I\left(y_i = c_j\right)}\right\} \tag{3.8}$$

in which α_k is the coefficient to find the nearest neighbors to the kth cluster center in the K-means approach.

3.2.5 A Toy Example

The ML-based clustering algorithm processes the historical data of EV users and divides their behaviors into three groups, as shown in Fig. 3.7. Three of the six features in the tuple are used for visualization: start time, end time, and weather correlation. The three possible categories can be cold weather, normal weather, and hot weather. User behavior is labeled using human intelligence following the clustering. According to the results of the simulation, it could be concluded that red- and blue-colored users have charged in cooler climates, indicating that their charging behavior is more regular and predictable. Therefore, such users will be able to collaborate in the centralized scheduling of the power network. In contrast, orange-colored users have more inefficient charging behaviors in hotter times.

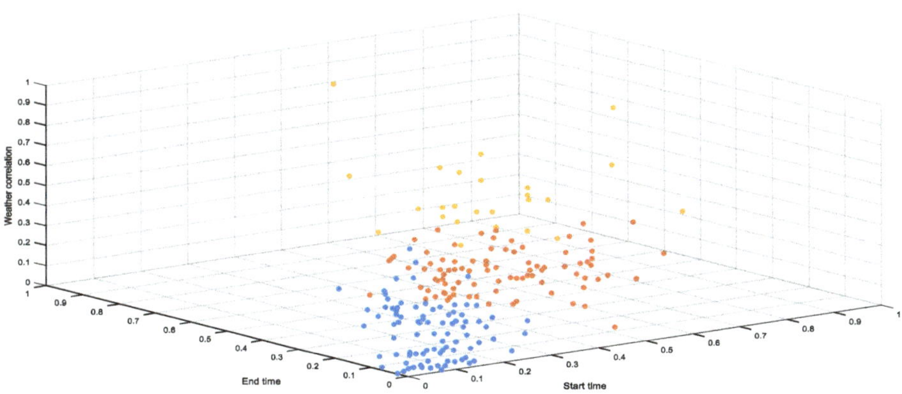

Fig. 3.7 EV user charging behaviors

Therefore, it is not helpful for these users to participate in the centralized scheduling of the power network.

3.2.6 Application of Charging Pattern Recognition in Smart Charging

Smart charging reduces EV-induced load peaks and flattens the load profile to make integration of RESs at the system levels and the local levels easier over shorter period scales. To be specific, adjusting and clustering the charging patterns of EVs parked in parking lots for the majority of time (90–95% of time) could help with peak shaving, ancillary services, and backup power [45].

Clustering approaches are also applied to EV smart charging in order to determine the most common load patterns. The EV charging clustering technique is utilized to find the clusters with identical charging items using a multivariate dataset. EVs can change their own charging patterns using smart charging to flatten peak load, fill up demand valleys, and assist the real-time balancing of the grid by altering charging and discharging levels. Furthermore, using EVs as a flexible source through smart charging technologies could lessen the requirement for investment in power plants that produce a vast amount of CO2 to balance fluctuations in renewable generations.

Furthermore, advancements in information and communications technology, such as data management and analytics from users, charging stations, and charging patterns could improve smart charging functionality and optimize grid service presentation.

3.3 Status Quo, Challenges, and Outlook

In this chapter, definitions of EVs and a general taxonomy of them were provided. Then, the EV integration methods (V1G, V2G, G2V, V2H, and V2B) with the electricity network were introduced. In the following, general challenges and problems of EV integration were introduced. However, data challenges of the high penetration of the EVs were investigated as a separate section. A general classification of EV data was also proposed in this section.

Moreover, various aspects of the EVs' smart charging energy management were discussed, such as concepts, applications, challenges, and opportunities. A framework for designing a smart charging system for integrating the EV-RES system into the grid was also presented. Next, some of the recent literature on EV integration and practical EV projects were examined to demonstrate the current situation of EV's employment around the globe.

Afterward, various benefits, reasons, and methods of EV charging pattern identification were presented in the second section. Then, general concepts and principles of clustering methods were discussed, and charging pattern clustering advantages were addressed. Next, different types of ML algorithms for clustering the charging patterns were presented and discussed in detail, including unsupervised learning and supervised learning.

Furthermore, different types of ML algorithms for clustering the charging patterns were presented and discussed in detail, including unsupervised learning and supervised learning. In addition, the proposed ML-based approach to cluster the EV charging behaviors was introduced, and the related formulations of the K-means and K-NN algorithms were presented. Also, a toy example was provided in order to demonstrate the practicality and effectiveness of the proposed method. Eventually, the application of charging pattern recognition in smart charging was also discussed briefly.

There are several research gaps that could be filled in future research. Some of the important future research outlooks from the authors' point of view are presented as follows:

- The lack of a labeled dataset posed a problem for this research. Further evaluation could be conducted using a properly labeled dataset to verify whether the promising results and training method are sound.
- Deep learning and reinforcement learning are subsets of ML that utilize artificial neural networks. These methods, unlike ML models, contain a large amount of composition of learned functions. Therefore, using deep and reinforcement learning, which could result in better efficiency, should be considered in future works.
- Further studies should be conducted to identify the applications of the clustering methods to analyze the behavior of drivers and the relevant aspects.

3.4 Concluding Remarks

The goal of this chapter was to look into and examine the charging behaviors of EVs. In this chapter, an unsupervised learning method, i.e., K-means, was utilized in order to cluster the charging patterns of the EVs. The main contribution of this chapter was to consider items that have not been appropriately considered in the previous literature. The most popular items for investigating the EV charging behaviors are start time, stay duration, and energy demand. Other items which might have significant effects on the classification of the EVs are traffic patterns and weather that have been addressed in this chapter. Finally, a supervised learning method, i.e., K-NN, was also utilized for further classification of the new dataset of the EVs.

References

1. Arasteh, H., Kia, M., Vahidinasab, V., Shafie-khah, M., & Catalão, J. P. S. (2020). Multiobjective generation and transmission expansion planning of renewable dominated power systems using stochastic normalized normal constraint. *International Journal of Electrical Power & Energy Systems, 121*. https://doi.org/10.1016/j.ijepes.2020.106098
2. Shekari, M., & Moghaddam, M. P. (2020). An introduction to blockchain-based concepts for demand response considering of electric vehicles and renewable energies. *2020 28th Iranian Conference on Electrical Engineering, 2020*. https://doi.org/10.1109/ICEE50131.2020.9260825
3. Pirouzi, S., Aghaei, J., Vahidinasab, V., Niknam, T., & Khodaei, A. (2018). Robust linear architecture for active/reactive power scheduling of EV integrated smart distribution networks. *Electric Power Systems Research, 155*, 8–20. https://doi.org/10.1016/j.epsr.2017.09.021
4. Wooding, B., Vahidinasab, V., & Soudjani, S. (2020). Formal controller synthesis for frequency regulation utilising electric vehicles. *SEST 2020 – 3rd International Conference on Smart Energy Systems and Technologies*. https://doi.org/10.1109/SEST48500.2020.9203234
5. Aghapour, R., Sepasian, M. S., Arasteh, H., Vahidinasab, V., & Catalão, J. P. S. (2020). Probabilistic planning of electric vehicles charging stations in an integrated electricity-transport system. *Electric Power Systems Research, 189*. https://doi.org/10.1016/j.epsr.2020.106698
6. Shahriar, S., Al-Ali, A. R., Osman, A. H., Dhou, S., & Nijim, M. (2020). Machine learning approaches for EV charging behavior: A review. *IEEE Access, 8*, 168980–168993. https://doi.org/10.1109/ACCESS.2020.3023388
7. National Grid. (2020). *Future energy scenarios navigation*. No. July, pp. 1–124.
8. Anastasiadis, A. G., Kondylis, G. P., Polyzakis, A., & Vokas, G. (2019). Effects of increased electric vehicles into a distribution network. *Energy Procedia, 157*, 586–593. https://doi.org/10.1016/j.egypro.2018.11.223
9. Huda, M., Tokimatsu, K., & Aziz, M. (2020). Techno economic analysis of vehicle to grid (V2G) integration as distributed energy resources in Indonesia power system. *Energies, 13*(5). https://doi.org/10.3390/en13051162
10. Sanguesa, J. A., Torres-Sanz, V., Garrido, P., Martinez, F. J., & Marquez-Barja, J. M. (2021). A review on electric vehicles: Technologies and challenges. *Smart Cities, 4*(1), 372–404. https://doi.org/10.3390/smartcities4010022
11. What are the different types of electric vehicles? *Nrma*, 2021. https://www.mynrma.com.au/cars-and-driving/electric-vehicles/buying/types-of-evs. Accessed November 15, 2021.
12. Coulter, J. (2019). BEV, EREV, PHEV, HEV – What do they mean? Here's your electric vehicle dictionary – Current EV blog. *Current EV Blog*. https://currentev.com/blog/bev-erev-phev-hev-what-do-they-mean-an-ev-dictionary/. Accessed November 15, 2021.
13. Plötz, P., Moll, C., Bieker, G., Mock, P., & Li, Y.. (2020) *Real-world usage of plug-in hybrid electric vehicles: Fuel consumption, electric driving, and CO2 emissions | International Council on Clean Transportation*. Washington, DC. [Online]. Available: https://theicct.org/publications/phev-real-world-usage-sept2020
14. Ehsani, M., Gao, Y., & Emadi, A. (2019). Hybrid electric vehicles. *Modern Eelectric, Hybrid Electric, and Fuel Cell Vehicles*, 1–20. https://doi.org/10.1201/9781420054002-5
15. Fuel Cell & Hydrogen Energy Association. (2015). Fuel cell electric vehicles. *U.S. Department of Energy – Energy Efficiency and Renewable Energy*. https://afdc.energy.gov/vehicles/fuel_cell.html. Accessed November 15, 2021.
16. Punter, M. (2021). EREVs: The extended range electric vehicle explained. *Electric For The Profile*. https://e4tp.com/the-extended-range-electric-vehicle-explained/. Accessed November 15, 2021.
17. Pasaoglu, G., Fiorello, D., Martino, A., Zani, L., Zubaryeva, A., & Thiel, C. (2014). Travel patterns and the potential use of electric cars – Results from a direct survey in six European countries. *Technological Forecasting and Social Change, 87*, 51–59. https://doi.org/10.1016/j.techfore.2013.10.018

18. Aziz, M., Oda, T., Mitani, T., Watanabe, Y., & Kashiwagi, T. (2015). Utilization of electric vehicles and their used batteries for peak-load shifting. *Energies, 8*(5), 3720–3738. https://doi.org/10.3390/en8053720
19. Mohammad, A., Zamora, R., & Lie, T. T. (2020). Integration of electric vehicles in the distribution network: A review of PV based electric vehicle modelling. *Energies, 13*(17). https://doi.org/10.3390/en13174541
20. Tushar, W., Yuen, C., Mohsenian-Rad, H., Saha, T., Poor, H. V., & Wood, K. L. (2018). Transforming energy networks via peer-to-peer energy trading: The potential of game-theoretic approaches. *IEEE Signal Processing Magazine, 35*(4), 90–111. https://doi.org/10.1109/MSP.2018.2818327
21. Rigas, E. S., Ramchurn, S. D., & Bassiliades, N. (2015). Managing electric vehicles in the smart grid using artificial intelligence: A survey. *IEEE Transactions on Intelligent Transportation Systems, 16*(4), 1619–1635. https://doi.org/10.1109/TITS.2014.2376873
22. Hou, L., Wang, C., & Yan, J. (2021). *Electric vehicle charging scheduling in green logistics: Challenges, approaches and opportunities*. p. 2 [Online]. Available: http://arxiv.org/abs/2103.07635.
23. Pareek, S., Sujil, A., Ratra, S., & Kumar, R. (2020). Electric vehicle charging station challenges and opportunities: A future perspective. *Proceedings – 2020 International Conference on Emerging Trends in Communication, Control and Computing ICONC3 2020*. https://doi.org/10.1109/ICONC345789.2020.9117473.
24. Di Paolo, M. (2018). Analysis of harmonic impact of electric vehicle charging on the electric power grid, based on smart grid regional demonstration project – Los Angeles. *2017 IEEE Green Energy and Smart Systems Conference IGESSC 2017, 2017-Novem*, 1–5. https://doi.org/10.1109/IGESC.2017.8283460
25. Monteiro, V., Gonçalves, H., & Afonso, J. L. (2011). Impact of electric vehicles on power quality in a Smart Grid context. *Proceeding of the International Conference on Electrical Power Quality and Utilisation EPQU*, 660–665. https://doi.org/10.1109/EPQU.2011.6128861
26. Varghese, A. S., Thomas, P., & Varghese, S. (2017). An efficient voltage control strategy for fast charging of plug-in electric vehicle. *2017 Innovations in Power and Advanced Computing Technologies i-PACT 2017, 2017-Janua*, 1–4. https://doi.org/10.1109/IPACT.2017.8245074
27. Teixeira, A. C. R., & Sodré, J. R. (2018). Impacts of replacement of engine powered vehicles by electric vehicles on energy consumption and CO2 emissions. *Transportation Research Part D: Transport and Environment, 59*, 375–384. https://doi.org/10.1016/j.trd.2018.01.004
28. Habib, S., Khan, M. M., Abbas, F., Sang, L., Shahid, M. U., & Tang, H. (2018). A comprehensive study of implemented international standards, technical challenges, impacts and prospects for electric vehicles. *IEEE Access, 6*, 13866–13890. https://doi.org/10.1109/ACCESS.2018.2812303
29. Ramanujam, A., Sankaranarayanan, P., Vasan, A., Jayaprakash, R., Sarangan, V., & Sivasubramaniam, A. (2017). *Quantifying the impact of electric vehicles on the electric grid* (pp. 228–233). https://doi.org/10.1145/3077839.3077854
30. García-Álvarez, J., González, M. A., & Vela, C. R. (2018). Metaheuristics for solving a real-world electric vehicle charging scheduling problem. *Applied Soft Computing Journal, 65*, 292–306. https://doi.org/10.1016/j.asoc.2018.01.010
31. Moghaddam, Z., Ahmad, I., Habibi, D., & Phung, Q. V. (2017). Smart charging strategy for electric vehicle charging stations. *IEEE Transactions on Transportation Electrification, 4*(1), 76–88. https://doi.org/10.1109/TTE.2017.2753403
32. Kuran, M. S., Carneiro Viana, A., Iannone, L., Kofman, D., Mermoud, G., & Vasseur, J. P. (2015). A smart parking lot management system for scheduling the recharging of electric vehicles. *IEEE Transactions on Smart Grid, 6*(6), 2942–2953. https://doi.org/10.1109/TSG.2015.2403287
33. Foley, A., Tyther, B., Calnan, P., & Gallachóir, B. Ó. (2013). Impacts of Electric Vehicle charging under electricity market operations. *Applied Energy, 101*, 93–102. https://doi.org/10.1016/j.apenergy.2012.06.052

34. Xie, R., Wei, W., Khodayar, M. E., Wang, J., & Mei, S. (2018). Planning fully renewable powered charging stations on highways: A data-driven robust optimization approach. *IEEE Transactions on Transportation Electrification, 4*(3), 817–830. https://doi.org/10.1109/TTE.2018.2849222
35. Zhang, T., Chen, W., Han, Z., & Cao, Z. (2014). Charging scheduling of electric vehicles with local renewable energy under uncertain electric vehicle arrival and grid power price. *IEEE Transactions on Vehicular Technology, 63*(6), 2600–2612. https://doi.org/10.1109/TVT.2013.2295591
36. Li, B., Kisacikoglu, M. C., Liu, C., Singh, N., & Erol-Kantarci, M. (2017). Big data analytics for electric vehicle integration in green smart cities. *IEEE Communications Magazine, 55*(11), 19–25. https://doi.org/10.1109/MCOM.2017.1700133
37. Wipro Insights. (2012). *Big data helps electric vehicles shift gears*. https://www.wipro.com/blogs/wipro-insights/big-data-helps-electric-vehicles-shift-gears/. Accessed July 13, 2021.
38. Mellor, C. (2020). Data storage estimates for intelligent vehicles vary widely. *Blocks and Files* [Online]. Available: https://blocksandfiles.com/2020/01/17/connected-car-data-storage-estimates-vary-widely/
39. Pevec, D., Babic, J., & Podobnik, V. (2019). Electric vehicles: A data science perspective review. *Electron, 8*(10). https://doi.org/10.3390/electronics8101190
40. Anastasiadis, A. G., et al. (2017). Economic benefits from the coordinated control of distributed energy resources and different charging technologies of electric vehicles in a smart microgrid. *Energy Procedia, 119*, 417–425. https://doi.org/10.1016/j.egypro.2017.07.125
41. Su, J., Lie, T. T., & Zamora, R. (2019). Modelling of large-scale electric vehicles charging demand: A New Zealand case study. *Electric Power Systems Research, 167*, 171–182. https://doi.org/10.1016/j.epsr.2018.10.030
42. Internation Renwable Energy Agency. (2019). *Electric-vehicle smart charging innovation landscape brief*. [Online]. Available: www.irena.org
43. Faddel, S., Al-Awami, A. T., & Mohammed, O. A. (2018). Charge control and operation of electric vehicles in power grids: A review. *Energies, 11*(4). https://doi.org/10.3390/en11040701
44. Frendo, O., Graf, J., Gaertner, N., & Stuckenschmidt, H. (2020). Data-driven smart charging for heterogeneous electric vehicle fleets. *Energy AI, 1*. https://doi.org/10.1016/j.egyai.2020.100007
45. IRENA. (2019). Innovation outlook: Smart charging for electric vehicles. *National Conference on Innovations in Electrical Power and Energy Systems (NCIEPES-19), M. Kumarasamy College of Engineering, Karur*, p. 138.
46. Kim, B. (2013). Smart charging architecture for between a plug-in electrical vehicle (PEV) and a smart home. *2013 International Conference on Connected Vehicles and Expo, ICCVE 2013 – Proceedings*, pp. 306–307. https://doi.org/10.1109/ICCVE.2013.6799811
47. Dimitrakopoulos, G. (2011). Intelligent transportation systems based on internet-connected vehicles: Fundamental research areas and challenges. *2011 11th International Conference on ITS Telecommunications (ITST 2011)*, pp. 145–151. https://doi.org/10.1109/ITST.2011.6060042
48. Tian, W., He, J., Niu, L., Zhang, W., Wang, X., & Bo, Z. (2012). Simulation of vehicle-to-grid (V2G) on power system frequency control. *2012 IEEE Innovative Smart Grid Technologies – Asia (ISGT Asia 2012)*. https://doi.org/10.1109/ISGT-Asia.2012.6303105
49. Hajar, K., Guo, B., Hably, A., & Bacha, S. (2021). *Smart charging impact on electric vehicles in presence of photovoltaics* (pp. 643–648). https://doi.org/10.1109/icit46573.2021.9453600
50. Frendo, O., Gaertner, N., & Stuckenschmidt, H. (2020). Improving smart charging prioritization by predicting electric vehicle departure time. *IEEE Transactions on Intelligent Transportation Systems*, 1–8. https://doi.org/10.1109/tits.2020.2988648
51. Babic, J., Carvalho, A., Ketter, W., & Podobnik, V. (2017). Evaluating policies for parking lots handling electric vehicles. *IEEE Access, 6*, 944–961. https://doi.org/10.1109/ACCESS.2017.2777098

52. Divyapriya, S., Amutha, & Vijayakumar, R. (2018). Design of residential plug-in electric vehicle charging station with time of use tariff and IoT technology. *ICSNS 2018 – Proceedings of IEEE International Conference on Soft-Computing and Network Security*https://doi.org/10.1109/ICSNS.2018.8573637
53. Kara, E. C., Macdonald, J. S., Black, D., Bérges, M., Hug, G., & Kiliccote, S. (2015). Estimating the benefits of electric vehicle smart charging at non-residential locations: A data-driven approach. *Applied Energy, 155*, 515–525. https://doi.org/10.1016/j.apenergy.2015.05.072
54. Cao, Y., et al. (2018). Mobile edge computing for big-data-enabled electric vehicle charging. *IEEE Communications Magazine, 56*(3), 150–156. https://doi.org/10.1109/MCOM.2018.1700210
55. Zhang, T., Pota, H., Chu, C. C., & Gadh, R. (2018). Real-time renewable energy incentive system for electric vehicles using prioritization and cryptocurrency. *Applied Energy, 226*, 582–594. https://doi.org/10.1016/j.apenergy.2018.06.025
56. Al-Ogaili, A. S., et al. (2019). Review on scheduling, clustering, and forecasting strategies for controlling electric vehicle charging: Challenges and recommendations. *IEEE Access, 7*, 128353–128371. https://doi.org/10.1109/ACCESS.2019.2939595
57. Fitzgerald, G., Nelder, C., & Newcomb, J. (2016). *Electric vehicles as distributed energy resources* (p. 78).
58. Sherwood, J., Chitkara, A., Cross-Call, D., & Li, B. (2016). *A review of alternative rate designs: Industry experience with time-based and demand charge rates for mass-market customers*. no. May.
59. ENTSO. (2021). Electric vehicle integration into power grids. *ENTSO-E Position Pap.*, no. March, [Online]. Available: https://eepublicdownloads.entsoe.eu/clean-documents/Publications/Position%20papers%20and%20reports/210331_Electric_Vehicles_integration.pdf
60. Engel, H., Hensley, R., Knupfer, S., & Sahdev, S.. (2018). The Potential impact of electric vehicles on global energy systems. *McKinsey Center for Future Mobility*, no. Exhibit 1, p. 8, [Online]. Available: https://www.mckinsey.com/industries/automotive-and-assembly/our-insights/the-potential-impact-of-electric-vehicles-on-global-energy-systems
61. Das, H. S., Rahman, M. M., Li, S., & Tan, C. W. (2020). Electric vehicles standards, charging infrastructure, and impact on grid integration: A technological review. *Renewable and Sustainable Energy Reviews, 120*. https://doi.org/10.1016/j.rser.2019.109618
62. Patil, H., & Kalkhambkar, V. N. (2021). Grid integration of electric vehicles for economic benefits: A review. *Journal of Modern Power Systems and Clean Energy, 9*(1), 13–26. https://doi.org/10.35833/MPCE.2019.000326
63. Tiwari, D., Sheikh, M. A. A., Moyalan, J., Sawant, M., Solanki, S. K., & Solanki, J. (2020). Vehicle-to-grid integration for enhancement of grid: A distributed resource allocation approach. *IEEE Access, 8*, 175948–175957. https://doi.org/10.1109/ACCESS.2020.3025170
64. Tan, K. M., Ramachandaramurthy, V. K., & Yong, J. Y. (2016). Integration of electric vehicles in smart grid: A review on vehicle to grid technologies and optimization techniques. *Renewable and Sustainable Energy Reviews, 53*, 720–732. https://doi.org/10.1016/j.rser.2015.09.012
65. Hu, X. (2018). Integration of EVs with a smart grid. *Modelling, Dynamics and Control of Electrified Vehicles*, 475–496. https://doi.org/10.1016/B978-0-12-812786-5.00014-8
66. Anthony, B., Jr. (2021). Integrating electric vehicles to achieve sustainable energy as a service business model in smart cities. *Frontiers in Sustainable Cities, 3*. https://doi.org/10.3389/frsc.2021.685716
67. SDG&E. (2014). SDG&E's electric vehicle grid integration pilot program. *San Diego Gas & Electric*. https://www.sdge.com/regulatory-filing/10676/sdge-electric-vehicle-grid-integration-pilot-program. Accessed July 20, 2021.
68. Nuvve Corporation. Vehicle-to-grid technology. *Nuvve Corporation*. https://nuvve.com/technology/. Accessed July 20, 2021.
69. Andersen, P. B., et al. (2017). *The Parker project*.
70. Pevec, D., Babic, J., Kayser, M. A., Carvalho, A., Ghiassi-Farrokhfal, Y., & Podobnik, V. (2018). A data-driven statistical approach for extending electric vehicle charging infrastructure.

International Journal of Energy Research, 42(9), 3102–3120. https://doi.org/10.1002/er.3978
71. Spoelstra, J. C. (2014). *Charging behaviour of Dutch EV drivers*. Utrecht University.
72. Bin Moon, H., Park, S. Y., Jeong, C., & Lee, J. (2018). Forecasting electricity demand of electric vehicles by analyzing consumers' charging patterns. *Transportation Research Part D: Transport and Environment, 62*, 64–79. https://doi.org/10.1016/j.trd.2018.02.009
73. Kim, Y., & Kim, S. (2021). Forecasting charging demand of electric vehicles using time-series models. *Energies, 14*(5). https://doi.org/10.3390/en14051487
74. Ul-Haq, A., Cecati, C., & El-Saadany, E. (2018). Probabilistic modeling of electric vehicle charging pattern in a residential distribution network. *Electric Power Systems Research, 157*, 126–133. https://doi.org/10.1016/j.epsr.2017.12.005
75. Hao, X., Wang, H., Lin, Z., & Ouyang, M. (2020). Seasonal effects on electric vehicle energy consumption and driving range: A case study on personal, taxi, and ridesharing vehicles. *Journal of Cleaner Production, 249*. https://doi.org/10.1016/j.jclepro.2019.119403
76. Verma, A., Asadi, A., Yang, K., & Tyagi, S. (2015). A data-driven approach to identify households with plug-in electrical vehicles (PEVs). *Applied Energy, 160*, 71–79. https://doi.org/10.1016/j.apenergy.2015.09.013
77. Mostafa, S. M., & Amano, H. (2019). Effect of clustering data in improving machine learning model accuracy. *Journal of Theoretical and Applied Information Technology, 97*(21), 2973–2981.
78. Straka, M., & Buzna, L. (2019). Clustering algorithms applied to usage related segments of electric vehicle charging stations. *Transportation Research Procedia, 40*, 1576–1582. https://doi.org/10.1016/j.trpro.2019.07.218
79. Crozier, C., Apostolopoulou, D., & McCulloch, M. (2018). Clustering of usage profiles for electric vehicle behaviour analysis. *Proceedings – 2018 IEEE PES Innovative Smart Grid Technologies Conference (ISGT-Europe 2018)*. https://doi.org/10.1109/ISGTEurope.2018.8571707
80. Miyazaki, K., Uchiba, T., & Tanaka, K. (2020). Clustering to predict electric vehicle behaviors using State of Charge data. *International Conference on Environment and Electrical Engineering and 2020 IEEE Industrial and Commercial Power Systems Europe (EEEIC/I CPS Eur. 2020)*. https://doi.org/10.1109/EEEIC/ICPSEurope49358.2020.9160675
81. Shen, Y., Fang, W., Ye, F., & Kadoch, M. (2020). EV charging behavior analysis using hybrid intelligence for 5G smart grid. *Electron, 9*(1). https://doi.org/10.3390/electronics9010080
82. Salicrú, M., Vives, S., & Zheng, T. (2009). Inferential clustering approach for microarray experiments with replicated measurements. *IEEE/ACM Transactions on Computational Biology and Bioinformatics, 6*(4), 594–604. https://doi.org/10.1109/TCBB.2008.106
83. Lee, J., An, M., Kim, Y., & Seo, J. I. (2021). Optimal allocation for electric vehicle charging stations. *Energies, 14*(18). https://doi.org/10.3390/en14185781
84. Barkost, P. H. (2020). *Detecting EV charging from hourly smart meter data*. [Online]. Available: https://munin.uit.no/handle/10037/19274
85. Wang, Y., Chen, Q., Hong, T., & Kang, C. (2019). Review of smart meter data analytics: Applications, methodologies, and challenges. *IEEE Transactions on Smart Grid, 10*(3), 3125–3148. https://doi.org/10.1109/TSG.2018.2818167
86. Hsu, D. (2015). Comparison of integrated clustering methods for accurate and stable prediction of building energy consumption data. *Applied Energy, 160*, 153–163. https://doi.org/10.1016/j.apenergy.2015.08.126
87. Alghamdi, T. A. (2018). Secure and energy efficient path optimization technique in wireless sensor networks using dh method. *IEEE Access, 6*, 53576–53582. https://doi.org/10.1109/ACCESS.2018.2865909
88. Dubey, A., & Santoso, S. (2015). Electric vehicle charging on residential distribution systems: Impacts and mitigations. *IEEE Access, 3*, 1871–1893. https://doi.org/10.1109/ACCESS.2015.2476996
89. Ramachandran, A., Balakrishna, A., Kundzicz, P., & Neti, A. (2018). Predicting electric vehicle charging station usage: Using machine learning to estimate individual station statistics from physical configurations of charging station networks. *arXiv*.

90. Kodinariya, T. M., & Makwana, P. R. (2013). Review on determining number of cluster in K-means clustering. *Journal of Advance Research in Computer Science and Management Studies, 1*(6), 90–95.
91. Botev, Z. I., Grotowski, J. F., & Kroese, D. P. (2010). Kernel density estimation via diffusion. *The Annals of Statistics, 38*(5), 2916–2957. https://doi.org/10.1214/10-AOS799
92. Awad, M., & Khanna, R. (2015). Support vector regression. In *Efficient learning machines* (pp. 67–80). Apress.
93. Brownlee, J. (2016). *Classification and regression trees for machine learning*. http://machinelearningmastery.com/classification-and-regression-trees-for-machine-learning/
94. Pryinankur, S. (2019). *K-nearest neighbor in machine learning*. https://www.knowledgehut.com/blog/data-science/knn-for-machine-learning
95. Skillicorn, D. B. (2012). Understanding high-dimensional spaces. *SpringerBriefs in Computer Science, 0*(9783642333972), 1–108. https://doi.org/10.1007/978-3-642-33398-9

Chapter 4
Cybersecurity and Data Privacy Issues of Electric Vehicles Smart Charging in Smart Microgrids

Raziye Aghapour, Mehdi Zeraati, Farkhondeh Jabari, Mohammadreza Sheibani, and Hamidreza Arasteh

4.1 Introduction

There has been an increasing shift toward the application of electric and hybrid vehicles in the transportation sector due to the environmental threat that stems from global warming and greenhouse gases emitted from the burning of fossil fuels. In this regard, several technologies such as fuel cell hybrid electric vehicles (EVs) [1], hybrid vehicles or combustion- engine electric-motor based vehicles [2, 3], and EVs running on battery power [4] have attracted more attention because of their near-zero-emission footprints, lower maintenance, cost-effectiveness, safe and easy driving modes, and reduced noise pollution. The short driving range of these vehicles is of vital importance for drivers [5]. Therefore, the manufacturers of EVs are implementing methods to produce batteries with higher density in terms of power and energy and are optimally designing the electronic devices used in the batteries [6]. At the same time, the infrastructure required for charging the batteries has captured the researchers' attention in the electric power industry. In order to facilitate the charging process of EVs, electric vehicle supply equipment (EVSE) has progressively been expanded [7]. In addition to transferring the power, EVSE serves as an interface between a power system and EVs to exchange data. Smart charging establishes a communication mechanism between the EVSE and power system, which supports the improvement in efficiency and the customization of

R. Aghapour
Department of Electrical Engineering, Abbaspour School of Engineering, Shahid Beheshti University, Tehran, Iran

M. Zeraati (✉) · F. Jabari · M. Sheibani · H. Arasteh
Power Systems Operation and Planning Research Department, Niroo Research Institute (NRI), Tehran, Iran
e-mail: mzeraati@nri.ac.ir; fjabbari@nri.ac.ir; msheibani@nri.ac.ir; harasteh@nri.ac.ir

V. Vahidinasab, B. Mohammadi-Ivatloo (eds.), *Electric Vehicle Integration via Smart Charging*, Green Energy and Technology,
https://doi.org/10.1007/978-3-031-05909-4_4

a charging schedule by monitoring and managing the electrical power [8]. For the optimal connection and control of electric vehicles in local power systems, smart charging protocols are designed to reduce the operational costs, balance the peak demand [9], and facilitate the integration of renewable energy sources. From another point of view, with the emergence of the Internet of things (IoT), advanced communication protocols, cloud computing, and data mining techniques, power distribution companies can control and manage electrical power generation and consumption [10, 11]. In smart grids, EVs are among the main drivers of the IoT [12]. Smart charging technologies made it possible to provide a bidirectional power flow between plug-in electric vehicles (PEVs) and the main energy system through the EVSE. In addition, as a part of the authentication process, both personal information and financial data are exchanged during smart charging. Nevertheless, despite abundant opportunities and merits coming with EVs, regular monitoring and data exchange between the EVs and the power system make the system susceptible to various security issues [13]. Thus, the safety and security of the operation of EVSE are essential for individuals, vehicles, and the infrastructure of power systems.

According to the ENISA Smart Grid Threat Landscape report [14], cyberattacks are evolving rapidly, as vertically integrated power systems undergo a transition to modern decentralized systems with the emergence of such active high-wattage loads as EVs [15]. Recently, identifying the electric vehicle ecosystem vulnerabilities has attracted increasing attention within academia. Cybersecurity, physical security, and the power system resiliency may be endangered if an adversary lunch attacks on the EV nexus [12]. The study conducted in [16] argues that automakers should incorporate technologies like message encryption and authentication during the design phase and manufacturing process. For example, highlighting the existing limitation of the controller area network (CAN) in electric vehicles, Ref. [17] recommends modifications to CAN that enhance its security level without incurring the cost of overloading central processing unit. This study suggests lightweight security techniques such as message encryption, authentication, and protection against attacks. The simulation results demonstrate the practicality of the technique for Texas Instruments TIVA-C in terms of meeting the limits of message length and time delay. Although such measures can be adopted to prevent potential cyberattacks, they solely cannot ensure the security of the system entirely. Guo et al. [11] proposed a systematic approach based on model predictive control to evaluate the vulnerability of a core control system of an automated electric vehicle. Then, index-based metrics are developed to reflect the impact of an attack (if an attack does not influence these metrics, it might be minor). Therefore, they can be used in identifying attacks and countermeasures. Leveraging the deep learning technique, Ref. [18] puts forward a model to identify and classify the denial of service (DoS) attacks in EVSE. The proposed model was implemented in Python, and the simulation results show that both deep neural network (DNN) and long short-term memory (LSTM) algorithm perform with an accuracy of above 99%. However, LSTM outperforms its counterpart, DNN, according to accuracy, precision, measure, and memory. Motivated by the proven superiority of LSTM, the authors extended their work to identify bypass attacks at the infrastructure layer [19]. The study steps ahead and assesses the impact of attacks on a PV-powered, stand-alone EVSE communicating via 5G with a centralized remote controller. It was concluded that the impact of an attack hinges on its intensity and targeted components. Enjoying features such as hashing, consensus mechanism, and smart contracts makes the blockchain an attractive candidate for establishing secure communication in an EV ecosystem [20]. An example of deploying blockchain for charging electric vehicles is presented in [21], where authors proposed a contract-based energy blockchain. According to permissioned blockchain technology, preselected EVs can inspect and share transactions' accounts before implementing an energy-efficient delegated Byzantine fault tolerance algorithm based on reputation so as to achieve consensus. The process of making decision on energy-trading contracts between EVs and aggregators is modeled by leveraging contract theory, where the preference of EV users in power consumption and aggregators' conditions are optimally met. Finally, the assignment of contracts and the allocation of energy is implemented based on a dynamic model. Despite the robust security features of blockchains, it is shown that, if 51% of blockchain nodes are compromised, an adversary can delete all nodes and manipulate consensus conditions [22]. According to [20], artificial intelligence has proved its performance in complicated pattern recognition and detecting anomalies in power systems. Thus, it can be combined in different layers of blockchain for presenting sustainable stand-alone ecosystems. Authors of [23] aim at addressing the scalability issue of blockchain by adopting a deep reinforcement learning approach, as the advancement of the Internet of vehicles (IoV) entails dealing with massive data. In the proposed framework, the block size and intervals are adjusted according to the system dynamics, thus maximizing the performance of the EV transaction while maintaining its security, latency, and decentralization.

Abovementioned studies focus on the execution of cyberattacks on individual EVs or EVSE. However, due to prevalent communications in the EV nexus and mobility of vehicles, cyberattacks on EVs can spread very fast as an EV carrying malware travels around the city and infects other equipment and EVs on its way. This can potentially lead to jeopardizing the power system, so the propagation of attacks is much more impactful in decentralized power systems [12]. The IoT-connected, high-wattage EVs establish a new foothold for weaponizing them in order to launch large-scale demand-side cyberattacks [24]. Acharya et al. [25] investigated the vulnerabilities of an artificial intelligent-based demand response (DR) program and developed a data-driven mechanism (so-called causative attack) that is able to inflict frequency instability on the system. In an attempt to intrude into the learning scheme, an adversary manipulates monetary incentives to alter DR customers' sensitivities to DR events. To estimate the impact of each DR attribute, Shapley value theorem, a game-theoretical approach, is deployed. The analysis of launching the attack on the New York University microgrid revealed how changes in monetary incentives lead to power curtailment and, subsequently, frequency excursion. A botnet, a network of compromised EVs and EVSE, can be created by an attacker to carry out cyberattacks. The impact of a load-altering botnet attack on both a distribution and transmission system is studied in [26]. An omniscient attacker seeks to cause voltage drop and line congestion in the power system by directing EVs to charge at certain EVSE at a specific time. Even though 33- and 39-bus IEEE distribution and transmission systems are adjusted to launch the attack, real data obtained from the Toronto parking authority are used to represent the mobility and availability of EVs in charging stations. Simulation results showed that increasing charging demand using an EV botnet not only can jeopardize voltage levels and power flows in the distribution system but also lead to outages in the transmission system. Similarly, in the study carried out in [27], a botnet of charging stations and cellular equipment is formed to mount an attack against the California power system in an effort to cause an underfrequency outage. It is shown that the botnet attack creates a spike in demand that produces marked fluctuations in the supply-demand balance. This drops the frequency to a point which triggers underfrequency alarms. The study demonstrates that it is only required to take control over roughly 21% of California's EV population to carry out such an attack.

There are numerous motivations for attacking an EV charging station, from electricity theft to causing a disturbance in a local power system. Some attacks may be severe, and malwares could be spread through charging stations network, which potentially have significant impacts on a local power system. For the safe and secure charging of EVs, most EV charging stations have implemented some formats of information security and have considered the cyber threats and vulnerabilities of the EVSE in the designing stage of these cyber-physical systems (CPSs). Three criteria for cybersecurity in designing a CPS are [28]:

1. The availability of charging stations for monitoring: The availability offers the capability to identify and prevent various attacks such as DoS.
2. The integrity of data and information: The integrity of data and information provides the ability to protect against unauthorized changes. Protection should be presented against manipulating stored data (in charging stations, central servers, or common devices) or exchanging data and information among different entities.
3. The confidentiality of data transferred: The secrecy is kept in the information transfer process between different entities and devices using the confidentiality criteria.

In addition to the above security requirements, there are some special security and privacy requirements in EV networks, including [29]:

- Identity privacy: It is essential that private information – namely, EV user's identity, location, availability, and payment information – be not shared with both other EV users and EVSE.
- Entity authentication: In order to restrict impersonate attacks, every EVSE is required to verify the legitimacy of its newly arrived EVs.
- Power consumption privacy: Power-consuming data (current and target state of charge) should be aggregated in the encrypted domain. Smart grid operators should not collect the real-time electricity usage data of EV users.

One promising application of IoT is in the intelligent transportation sector. Intelligent transportation is referred to the application of modern technologies and strategic management in the transportation system. Such technologies as sensors and actuators can be effective in the enhancement of users' awareness of traffic network's condition and in the dynamic regulation and management of the traffic as well as the efficiency of transportation service through collecting and analyzing data with the help of wireless network communication.

Intelligent transportation systems (ITSs), as one of the main components of smart grids, have a substantial impact on cost reduction for charging EVs batteries, traffic control, and the improvement in users' safety. Modern automobiles are a CPS in which electrical-mechanical components are controlled with millions of lines of software codes. Vehicle components are controlled with a variety of electric control units (ECUs), which are linked to one another via an in-vehicular network connected to the Internet. Each ECU connected to the World Wide Web is prone to get attacked by malware. Simple systems without cybersecurity are the prime target for hackers to penetrate and access local network information. In EVs, all devices, even the simplest one, are controlled by ECUs, connected to the Internet via an in-vehicular network or via drivers' mobile phones in a wireless way. A hacker can access the vehicle network with the driver's mobile phone and take complete control of the vehicle by installing malware on each ECU [11]. Since EVs are equipped with modern communication, the probability of a cyber- and physical attack on vehicles rises. Therefore, communication interfaces of vehicles require a certain level of security against cyberattacks. In a physical attack, an attacker attempts to collect data or inject misleading information by directly connecting a USB port to an on-board diagnostic (OBD) port. In remote and wireless cyberattacks, an attacker does not need physical access to vehicles and connects them through Bluetooth and via telematics in a wireless and cellular way. Some of the primary purposes for carrying out cyberattacks on EVs are traffic control, opening desired routes, and vehicle theft [30].

Cyberattacks on EVs make it possible for an attacker to control the vehicles, resulting in devastating impacts such as brake disabling, turning headlights off, and steering manipulation. Spoofing attacks on Global Positioning System (GPS) can lead to deviation from routes. Due to the connection of battery charging infrastructure to power systems in charging stations, it is crucial to consider the cyber-physical security of connected and automated electric vehicles (CAEVs) in the optimal management and control of smart microgrids. Connections among CAEVs, charging stations, and smart grids can be exposed to cyberattacks. The absence of security monitoring puts CAEVs prone to a variety of cyberattacks like DoS and man-in-the-middle (MiTM), resulting in disruption to vehicle performance, as well as physical damages to the vehicle, people, and the system (infrastructure) [10]. Owning to the constant monitor and transfer of data between EVs and distribution companies, distribution systems are vulnerable to cyberattacks [31]. For example, in 2019, a cyberattack was reported by NERC in which a loophole in the data exchange bed of a firewall vendor had been exploited, resulting in DoS on the control center and several remote generation sites [32]. This issue was related to charging stations of EVs because of their high accessibility and power demand. Such attacks are more prevalent on EVs' chargers that suffer from adequate data privacy and cyber resiliency [13].

A cyberattack can be carried out on each element of a power and electric transportation system (PETS), including a meter, a phasor measurement unit (PMU), an EV, or a relay. Contagious attacks spread rapidly through PETS and infect electrical components, the computers of distribution companies, and servers of system operators. With a large number of interconnected devices in electric systems, cyberattacks can impact enormous equipment. As a result, a variety of automation and communication devices are employed to measure, monitor, and control power systems elements [33]. Therefore, ensuring their health and safety is a major challenge. The combination of information and communication technologies (ICT) makes cyberattacks more penetrable. Additionally, if EVs are adopted in a large scale, the transportation system is dependent on access to the power system. Power outage in the system can influence electrical transportation, or malware can slip from an EV to charging infrastructures and then circulate among power system components and distribution companies' servers [34].

The concept of cyber-physical security of the smart power systems is depicted in Fig. 4.1. The smart grid consists of IoT technology-based resources – such as photovoltaic panels, storage units, and flexible and shiftable loads, like EVs and heating, ventilation, and air conditioning (HVAC) – and communication infrastructure enabling the system to coordinate these resources with other components of the system. The smart energy networks are generally composed of the advanced metering infrastructure (AMI), PMUs, remote terminal units (RTUs), smart meters, power quality monitoring devices, programmable logic controller (PLC), and intelligent electronic devices (IEDs) for measuring the power system operational state variables, such as bus voltage, line current, active/reactive power flow, and frequency. After being collected through the data concentrators, these measured quantities communicate with a control center such as supervisory control and data acquisition (SCADA) or wide-area monitoring and control system (WAMS). Data exchange between different power system components, such as measurement units and SCADA/WAMS (cyber layer), is exposed to cyberattacks. In the physical layer, systematic attacks may occur in the generation, transmission, distribution, and consumer levels [35].

According to Fig. 4.1, in the cybersecurity analysis of charging EVs, the physical layer evaluates EV components from an electrical engineering perspective, whereas the cyber layer involves the communication interfaces connecting EVs to the system. These layers are described in details as follows:

Physical Layer

EVs are classified into three main groups, namely, battery electric vehicles (BEVs), plug-in hybrid electric vehicles (PHEVs), and hybrid electric vehicles (HEVs), according to their power sources. BEVs are powered through an electrochemical battery that is charged in either a residential or commercial EVSE from a local power system. PEVs are supplied with both a fuel-based internal combustion engine

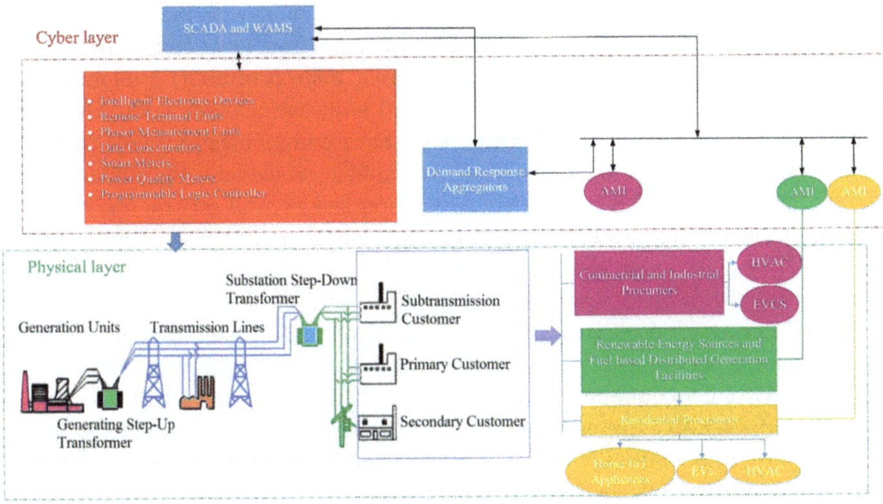

Fig. 4.1 The cyber-physical security of the smart power systems [35]

and an electrochemical battery [36]. HEVs are analogous to PHEVs, apart from that they connect to the system to exchange energy. The main parts of an EV are [37]:

- Battery: batteries work at a voltage of 300–700 [V] and are charged through a charger from a power system, an internal combustion engine, regenerative braking, or a combination of these methods [10]. Lithium-ion batteries are the most common type of batteries used in EVs, owning to their lower price and higher energy density.
- Power conditioning unit: EVs' batteries can be charged either directly through a residential or commercial charger or indirectly through an internal on-board charger. Both chargers use an AC/DC converter to transform received AC power from the system into battery-charger DC power.
- Motor and loads: an AC traction motor is a three-phase permanent magnet synchronous machine or an induction motor that works at a voltage of 240 [V]. EV loads include a three-phase compressor, ventilation fans, and USB chargers. On some occasions, such as when there is a radical difference between ambient temperature and that of the vehicle indoor, these loads can reduce the range of the EVs.

Cyber Layer

Generally, the cyber threats concerning EVs also include the EVSE, portable Internet service, road infrastructure, radio stations, other vehicles, and the original equipment manufacturers (OEMs). Advances in smart vehicle technologies have made it possible for in-vehicular networks to interact with external networks. Despite their intercommunication, these layers are different in imposed cyberattacks and in the spread patterns of cyberattacks inside and outside the EVs. EVs are equipped with ECUs, enabling the control of major functions like brake control or a battery managing center. Each ECU has a microprocessor, a memory, and input and output connectors. ECUs are linked with one another via CANs, local interconnected networks, and FlexRay. EVs have communication with an EVSE regarding charging process details. The communication includes the authentication of the signals of the EVs and EVSE, protection and control orders, and requisite software packages for the charging process. In addition, to establish communication with CAN bus, EVs have OBD ports for passing messages about the performance condition of the vehicle, speed, traffic patterns, and the battery condition. Such portable Internet services, as smartphone applications and web services, are used to gain remote access to EV characteristics, as well as monitor and control their charging. Also, EVs communicate with their manufacturers and OEM over a wide area network (WAN) – like cellular network and radio frequency – in order to update regular and ad hoc software. Modern EVs create communication links with road infrastructure to receive the signals of the traffic and other EVs through cellular blocks to enjoy a safe and relaxed driving experience [37].

4.2 Cyberattacks and Security Issues of EVs

Various types of cyberattacks and security challenges in different sectors impacting the EVs, namely, charging stations, communication, and battery charging systems, are introduced in this section. The purpose of this section is to introduce different to-EV-related sectors, which are vulnerable to cyberattacks. At first, different types of attacks on EVs are classified and introduced in Sect. 4.2.1. Then, since EV charging stations serve as power supply infrastructure for EVs, the cyber vulnerabilities and security bottlenecks in this sector are examined in Sect. 4.2.2. Section 4.2.3 is allocated to introduce cybersecurity challenges in the communication system of an EV nexus. Eventually, in Section 4.2.4, various kinds of cyberattacks on available power electronic hardware in an EV charger and on OBC are presented to highlight the vulnerability of the internal charging system of EVs and their related control circuits. Figure 4.2 is a graphical presentation of the main concepts that are investigated in this chapter.

4.2.1 Various Attacks on EVs

In this section, cyberattacks that may be launched on different components of EVs, namely, control systems, driving systems, and vehicle-to-everything (V2X) communication, are examined [38].

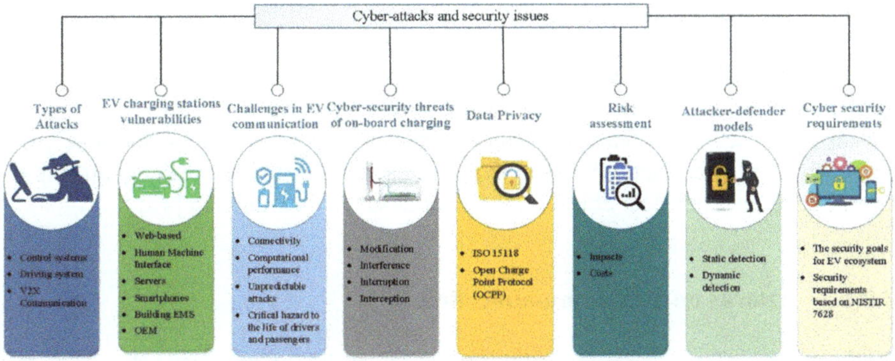

Fig. 4.2 Graphical presentation of the main concepts of the chapter

4.2.1.1 Attacks on Control Systems

Attacks on control systems usually target ECUs, the internal network of an EV, and car switches. Such attacks allow an attacker to gain unauthorized access to main ECUs, such as remote control, door lock receivers, and telematics devices [39]. An attacker can manipulate the car speed and represent misleading information on the car screen to a driver.

The process of vehicle hacking can be divided into three stages [40]:

1. The first stage is to compromise remotely.
2. The second stage is to send a message with cyber or physical components.
3. The last stage is to compel an ECU to perform the attacker's desired tasks.

Different methods of hacking include indirect physical access, short-board or long-board wireless connection, and attacks based on a dongle. The threats of hacking ECUs comprise collecting data and eavesdropping. Therefore, it is essential to constantly monitor and identify suspicious activities in order to develop a secure system for EVs.

Preliminary studies in internal network attacks have concerned such internal networks as CAN and FlexRay [41]. Some DoS attacks may be carried out on vehicles connecting to another one through CAN. These attacks depend upon the vulnerability of CAN protocol which is prone to attacks irrespective of manufacturers.

Attack scenarios on vehicles are described in the study done by Pan [42]. Additionally, threats that an attack poses on a vehicle by a CAN bus are introduced. A wrongdoer can attack a CAN bus, put a vehicle control system out of work, and inflict physical injuries on a diver or passengers.

Passive keyless entry and start (PKES) systems being used extensively in modern vehicles may be vulnerable to wireless attacks mounted by transceivers.

4.2.1.2 Attacks on Driving System Parts

Cyberattacks on an EV's driving system usually occur in sensors and mobile applications. Light detection and ranging, wireless access, autopilot and navigator, sensors, and actuator controllers can be investigated as vehicle vulnerable parts [43]. Also, hardware and operating system attacks, information leakage attacks by overhearing, and physical chip-tampering attacks are examined in [43].

4.2.1.3 Attacks on V2X Communication

V2X communication provides the ability to exchange messages between the vehicles and infrastructure. Vehicle-to-vehicle (V2V) and vehicle-to-infrastructure (V2I) are two kinds of the V2X communication [44]. V2V includes the exchanging of information, such as speed, heading, and brake status, among vehicles with on-board units. V2I facilitates the communication between the vehicles and roadside units. Vulnerabilities of theses communication structures can be targeted by cyberattackers and cause problems to EVs.

4.2.2 The Vulnerability of EV Charging Stations to Cyberattacks

From a cybersecurity perspective, attackers can commit security breaches using physical access points in an EV charging station enclosure and then through quasi-physical radio-frequency identification (RFID) technologies.

The vulnerability of charging systems of EVs to cyberattacks allows an attacker to take over complete control of them and have unauthorized access. Cyberattacks on EV charging stations include data tampering, information disclosure, and DoS. Similarly, applications working on the web and smartphones are of vital importance, since they allow a wrongdoer to acquire saved usernames, thus circumventing the authentication of EVs [35].

4.2.2.1 Web-Based Vulnerabilities

In a public charging station, more than one EVSE is likely to be connected to the power system. An attacker is capable of targeting all equipment connected to the system. Hence, there are three common vulnerability models concerning EVSE in this regard [45]:

- Vital vulnerabilities allowing an attacker to completely exploit the access to the Internet medium of an EVSE. Issuing commands to control the charging process is an example of such sabotage.

- High-risk vulnerabilities such as code injection that enable attackers to manage an operating system remotely.
- Moderate-risk vulnerabilities denote a situation that an attacker can disable the entering permission and achieve their malicious desires.

4.2.2.2 Human-Machine Interface Vulnerabilities and Physical Access Points

Public EV charging stations are generally not resilient enough against physical disruptions. Serial ports, USBs, Ethernet, magnetic card readers, and touching screens are at an attacker's disposal for physical penetration [46]. USB ports are suitable accessing points for cyberattacks, since they allow attackers to copy, correct, and delete stored data in the server of an EV charging station, including the information of recharged vehicles [46]. Furthermore, attackers are able to revise copied data and upload them anew. They also have the ability to skim the information of magnetic cards used for payment [47].

In addition to attacks targeting technical vulnerabilities, there are other physical attacks that can block users' access, disabling them from charging their vehicles. Natural and deliberate disasters, physical devastations or explosions, and the manipulation of cables connected to the station are examples of these cases.

4.2.2.3 The Vulnerability of Servers

Although communication protocols between EVs and charging station servers have not been standardized at a global level so far, open charge point protocol (OCPP) is being used in most charging stations. This protocol is structured according to client/server architecture, through which two sides can request information exchange. Nevertheless, this protocol is prone to cyberattacks on data privacy, message authenticity, and message integrity owing to the lack of client/server permits and end-to-end encryption [48]. These vulnerabilities give rise to the possibility of the theft, alteration, and forgery of EV charging data [49].

4.2.2.4 The Vulnerability of Smartphones

Both web-based and smartphone applications are essential for charging EVs at commercial charging stations. A growing number of these applications are being developed by charging station operators and other beneficiaries [24]. They are employed to determine public EV charging station locations, give permission to EVs to charge at charging stations, remotely control charging requests, and pay for charging services. Smartphone malware with unintended and undetected bugs can be harmful to EV charging. Some of the vulnerability of smartphone applications are reported to be under the threat of being exploited, leading to issuing unauthorized user permission and tampering with EV charging data, as well as charging settings. These attacks could potentially cause damage to charging stations and EVs connected to them.

For example, the authentication procedure of EV users in an EVSE is usually established with either an RFID label or smartphone applications. A smartphone application is open to the World Wide Web that can increase the scale of the attack. Furthermore, when the only way to validate users is RFID labels, this can expose additional vulnerabilities. Some examples to be mentioned include reverse engineering using stealing information on a chip, eavesdropping due to unencrypted data exchange in RFID systems, and DoS through making interruption in communication between a label and a reader by emitting radio jamming [50].

4.2.2.5 The Vulnerability of Building Energy Management System and Grid Interface

Direct connection between an EV charging station and an electrical company is standardized by protocols such as OpenADR, which uses data encryption and digital signature through a WAN. In a building, building energy management system (BEMS) coordinates the performance of a charging station with a variety of intelligent equipment and sensors – like television, heating system, HVAC, and security cameras – via infrastructures such as WiFi, ZigBee, WiMAX, etc. Due to the absence of customer-side security measures, the probability of the intrusion of an attacker into BEMS increases. The remote control of smart home appliances over smartphone applications and the scatter supply chain of equipment and applications can serve as access points for launching attacks on a BEMS [35].

4.2.2.6 The Vulnerability of Original Equipment Manufacturers/Vendors

Making online provision of security patches and software updates put EV charging stations prone to vulnerabilities similar to that of EVs reported in [46]. The supply chain of EV charging stations, in-work communication channels for updating information, and internal networks of charging stations can be exploited to jeopardize updating software and security patches.

4.2.3 Cybersecurity Challenges in EV Communication

Vulnerabilities regarding EV communication can be divided into four challenges elaborated hereunder [51].

4.2.3.1 Limited Connectivity

While the connectivity of external links of vehicles is increasing, most of them are not able to update their software via over-the-air (OTA) updates, disallowing cars to be protected against the latest cyberattacks. Even if OTA updates become more standardized, vehicles will still be at the risk of improper function due to incomplete updates.

4.2.3.2 Limited Computational Performance

The computational performance of an EV is limited compared to that of a computer. Such limitations stem from the fact that EVs have greater longevity and should withstand higher temperature and high-frequency vibration in comparison to an average PC or a laptop. As a result, high computational deficiency enhances the probability of vehicle hacking compared to computer hacking. The limited computational performance also means that certain solutions to the cybersecurity of vehicles entail an exorbitant cost for implementation.

4.2.3.3 The Scenarios and Threats of Unpredictable Attacks

The architecture of a vehicle can be broken through different entry points, such as a vehicle database and remote communication technologies. New attacks are constantly being mounted, so it will be difficult for automakers to forecast the target of the upcoming attacks from hackers.

4.2.3.4 Critical Hazard to the Life of Drivers and Passengers

If merely a few sensors report wrong information or only a few incorrect messages are sent, a fault could occur in vehicle performance, putting the life of passengers and drivers in danger.

4.2.4 Data Privacy Challenges in Smart EV Networks

In current charging networks, industry-standard protocols – ISO 15118 and OCPP – establish communication between EVs and EVSEs [52]. However, as stated in Sect. 4.2.2.3, these industry-standard protocols are exposed to severe security and privacy menaces. There exists a trusted intermediate in the ISO 15118, storing the EV users' private information to validate and identify them [53]. Furthermore, in order to route vehicles to a proper EVSE, this intermediate tracks mobile EVs, arising serious challenges in case of either accidental or purposeful information leakage. In a similar fashion, in the OCPP charging protocol, a trusted central entity exchanges private data between EV users and EVSE. Therefore, both protocols are prone to impersonation attacks where an adversary can pretend to be the central entity or an EVSE, collecting EV users' personal data [52]. For example, by predicting an EV user's working hour pattern through their charging behavior, an attacker can easily launch such criminal activities as theft in the EV owner's houses when they are at work [29].

4.2.5 Classifying the Cybersecurity Threats of On-Board Charging

In a commercial charging station, the physical connections of an OBC not only transfer power but also facilitate data exchange between a vehicle and a charger. Therefore, any data alteration can severely affect the charging process. Cybersecurity threats concerning OBC can generally be categorized into four main groups: modification, interference, interruption, and interception. The purpose of an attacker is to manipulate control parameters. In so doing, the control server will become vulnerable to the changes and fluctuations in a system, leading to impinging upon the safe and secure operation of the system. Each of the cyberattacks on the OBC system of an EV is described as follows [13].

4.2.5.1 Modification

This attack is carried out with the purpose of destroying the integrity of information, so an unauthorized agent gains access to measured data. For example, an adversary might attempt to alter controllers, parameters, or sensors, potentially resulting in deteriorated performance of the system and undesired fluctuations in the regular performance of the system [54]. An attacker can do this by either increasing or decreasing the level of a signal. If voltage signals are manipulated by multiplying a gain factor, controllers will track the incorrect reference point. Depending on the decrease or increase in either sensor signals or the reference point, this attack may cause overvoltage damage to semiconductors, switches and the battery or diminish the performance of the controller.

4.2.5.2 Interference

In an interference attack, tampering with the set control or measured data is done in a way that eventually affects the dynamics of a system. In doing so, controllers may track manipulated waveform rather than process the fault in order to follow the set point. Thus, the dynamic performance of the system would be affected [55].

Injecting high-frequency white noise (above 1 kHz) into measured signals is one of these methods.

4.2.5.3 Interruption

Interruption refers to threats that aim at information availability. In such attacks, a troublemaker attempts to hamper the performance of the system by putting some parts of the system out of work. For instance, an attacker may deactivate the OBC system by halting control pulses from a central processor. DoS attacks in which the extraction and process of data are ceased by blocking the connection links fall into this group [54].

4.2.5.4 Interception

Interception attacks aim at the reliability of information and provide an attacker with full access to a cyber-equipment. Wiretapping, keystroke logging, and fiber tapping are instances of these attacks. Such attacks potentially expose the OBC to hazards like overvoltage, resulting in getting out of synchronization and the improper performance of the control system [54].

4.2.6 Risk Assessment

Risk assessment is essential to evaluate the economic consequences of the risk of a potential attack. In cyberattacks on an EV network, two factors of impact and cost are of considerable importance. The impact refers to elements that affect the chance and probability of a successful attack. The cost concerns the necessary resources that an attacker needs to achieve success. With considering both factors, the risk is defined as the ratio of the impact to the cost [56].

The impact can be quantified with numbers – 0, 1, 3, or 9 – representing the intensity of a specific breakdown scenario: 0 refers to the least serious scenario, and 9 indicates the most serious one. Similarly, the cost can have values 0.1, 1, 3, or 9. The higher risk number signifies the high-risk systems. Therefore, the risk is equal to the associated number to the impact divided by the number assigned for the cost, because the probability of the impact of an attack is directly proportional to the required resources to carry out that attack. In other words, the more the potential impact of an attack, the more the necessary resources. Such a scoring mechanism allows system operators to rank recognized risks to identify highly risky areas and prioritize mitigation efforts and efficiently allocate the security resources. Potential parameters for each risk-assessment variable will be introduced at the following. The distance between the values of impact and cost facilitates the understanding of different levels of risks, the impact, and the cost for a user.

Impact numbers represent the intensity of the impact. For example, the impact score may be assigned as follows:

- 0: stealing of the registration ID of an EV
- 1: destroying EVSE leading to charging disability
- 3: information disclosure in transferring data between vehicles and charging facilities
- 9: the spread of virus between EVs and chargers

Also, the cost numbers serve as the indicators for expenses and hurdles that an attacker should incur to launch the breakdown scenario. For instance, the cost scores can be assigned as the following:

- 0.1: malware causing the discharge of EVs in a system
- 1: EV charging data disclosure using charging station database
- 3: malware causing the overcharge and explosion of batteries
- 9: simultaneous fast charging leading to the overloading of a charging station transformer

These scores are established through a survey of experienced researchers on the required resources to carry out cyberattacks [56]. In both cases, the severity of a score increases with a rise in numbers. Therefore, a greater risk ratio indicates a higher level of urgency and exigency of an attack. In order to preserve an acceptable level of risk for a system, higher risks should be mitigated. Additionally, after assessing the risk of an identified attack, system operators should adopt a proper security solution [57]. Therefore, in the next section, the recognition of a cyberattack in a system is investigated. Then a model is introduced for defending against the identified security threats.

4.2.7 *The Review of Attacker-Defender Models*

Developing intrusion detection models are necessary to effectively recognize and monitor cyberattacks [57]. In the study conducted in [31], algorithms are developed for the identification of cyberattacks that can potentially impact PEV battery packs during the period of charging. Two algorithms are discussed: first, the static detection, which only uses measured variables, and then dynamic detection, which deploys the dynamics of the system accompanying measurements.

Consider a scenario in which a PEV is plugged into a power supply. The PEV is exposed to a cyberattack, in which an attacker has access to the PEV control system. In such a condition, the following concerns might be raised [31]:

- The attacker may meddle with the PEV control system in a way to overcharge batteries, resulting in devastating consequences, like an overheat-triggered explosion.

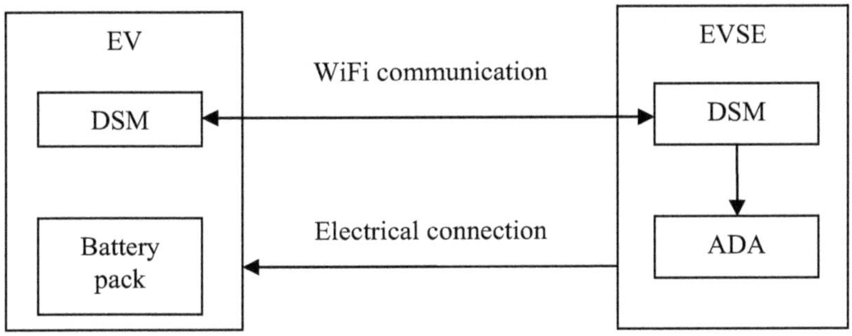

Fig. 4.3 The DSM framework [31]

- An attacker may prevent a PEV from charging by denial of charging attack, leading to the disconnection of vehicles and the instability of the power system, when there are a large number of plugged EVs.
- An attacker might want to gain access to the power system through an under-attack PEV.

The aim of the study conducted in [31] is to identify EVs which are vulnerable to the first two aforementioned security concerns. The cybersecurity problem of PEVs is formulated based on the diagnostic security module (DSM) framework. A DSM framework is depicted in Fig. 4.3. The considered PEV is plugged into an EVSE to be recharged. Equipped with the DSM framework, both the PEV and EVSE can connect to each other over WiFi communications. Furthermore, EVSE includes an attack detection algorithm which is embedded in its control system.

A static detector only uses output information (terminal voltage) measured in real time, whereas a dynamic detector takes advantage of the knowledge of system dynamics accompanying real-time measured data. The capabilities of both the static and dynamic detectors are also analyzed in this reference. Eventually, it is demonstrated that the dynamic detectors are able to better improve the cyberattack recognition capabilities compared to the static detectors [31].

After identifying the occurrence of an attack and assessing the risk associated to perceived threat, system operators should take remediation measures to restore the system. Based on the estimated threat level, one possible solution might be the temporary isolation of the comprised EVSE. The minimum number of EVs that should be disconnected to maintain an acceptable service level is modeled using mixed integer linear programming (MILP) in [58]. The objective function of the model (Eq. 4.1) is to minimize the utmost level of threats to all the connected EVSE, when an attack is recognized:

$$Z = \min_{x}\max_{j} \left(\theta_j \left(t_d\right) \times x_j\right) \forall j \in Y \tag{4.1}$$

In the above equation, Z denotes utmost level of threat to an EV charging device that should be minimized, j is the indicator for the number of EVSE, θ_j is the probability of the exposure of the jth EVSE to threats, x_j is a binary decision variable (1, if the jth EVSE is connected to the system; 0, otherwise), t_d is the time of the detection of the attack, and Y denotes the set of EVSEs. The level of threat at the time of recognition is defined according to Eq. 4.2:

$$\theta_j(t) = 1 - \left[1 - \theta_j\,(\mathrm{t} - 1)\right] \times \prod \left(1 - \beta \times \lambda_j\,(t - 1)\right) \tag{4.2}$$

where λ_j indicates the possibility of exposure to threats for the EV and β is the propagation probability of the cyberattack, transmitting from the EVSE to the EV and vice versa.

In Eq. 4.2, the first bracket represents the probability that an EV is not vulnerable to threats before the connection of new EVs. The second bracket indicates the likelihood that the attack on EVSE does not spread after the connection of new EVs.

By substitution of ω_j for $\theta_j(t_d) \times x_j$, the constraints of the model are formulated as follows:

$$\omega_j \leq x_j \tag{4.3}$$

$$\omega_j \leq \theta_j\,(t_d) \tag{4.4}$$

$$\omega_j \geq \theta_j\,(t_d) - \left(1 - x_j\right) \tag{4.5}$$

$$\omega_\mathrm{j} \geq 0 \tag{4.6}$$

$$\omega_j \leq Y \tag{4.7}$$

Equation 4.8 is added to the model to ensure that the connection of EV charging devices is maintained with threat levels below a specific threshold:

$$\theta_j\,(t_d) > T_j - x_j \forall j \in Y \tag{4.8}$$

where T_j is the threshold value for an EV infrastructure. The disconnection of EVSE may affect the power supply to EVs. As a result, a constraint should be introduced to make sure of meeting the power demand, as:

$$\sum_{j \in Y} C_j x_j \geq D_{\mathrm{EV}} \tag{4.9}$$

where C_j is the capacity of the jth EVSE and D_{EV} is the predicted demand of EV charging stations at the time of t_d.

4.2.8 *Cybersecurity Requirements*

4.2.8.1 The Security Goals for EV Ecosystem

For the management of charging times, an EV ecosystem highly relies on safe message exchanges. The integrity and authenticity of a message are the main criteria. Although the availability criterion is also important, it can be violated for few seconds. Confidentiality is crucial for such messages, as financial information and drivers' privacy [57].

In addition, machine-to-machine communications (M2M), which are expected to become the most prevalent method of communication in smart grids, require the heightened security of device identity.

All security architectures for an EV ecosystem should meet the following security targets:

- Authentic information with provenance confirmation for charging management
- The processing of safe payment to achieve the integrity/authenticity of exchange, the affirmation of customers, and privacy
- Safe integrity with power systems' information network, requiring highly secured equipment in an EV ecosystem and a safe integrating medium for controlled information flow
- The strong equipment confirmation to guarantee the safe M2M communications

These goals are central to the protection of system operators' revenue and the preservation of the system stability. It is particularly important to ensure equipment identity due to the following reasons: First, EVs may have much more power demands compared to the traditional households. The secure identification of equipment is essential to correspond to the increased risk of these large charging loads. Also, in order to attain demand-side management, an EV should authenticate thoroughly. Second, for the safe M2M communications, equipment should have the ability to automatically authenticate one another without human intervention. Third, to support V2G, the registration of battery characteristics is estimated to be vitally important [15].

4.2.8.2 Security Requirements Based on NISTIR 7628

NISTIR 7628 [59] (that contains a set of instructions and guidelines for the cybersecurity of smart grids) and ENISA [60] are introduced here that could be used to identify the security requirements of an EV ecosystem. Both of these documents have similar instructions; however, ENISA has guidelines for three different levels of cybersecurity, so as to be compatible with various architectures.

NISTIR 7628 identifies 46 agents, divided into 7 operating sectors: mass production, transmission, distribution, costumers, markets, operators, and service providers. More than 130 logic interfaces media are recognized among these agents to represent main features for different kinds of information exchange. Logical interfaces which have similar security requirements are classified into 22 groups, each of which has a distinct set of security requirements and priority values (high, medium, low) for security goals of confidentiality, integrity, and availability.

4.3 Status Quo, Challenges, and Outlook

EVs have become one of the tools to solve environmental challenges due to their impacts on reducing greenhouse gas emissions. In the past few years, various countries have been looking for ways to increase the presence of these vehicles in the transportation system. A number of countries have also been able to achieve this goal by establishing the necessary infrastructure.

Also, the presence of EVs poses some challenges. One of these challenges is the possibility of cyberattacks. Security challenges and cyberattacks may happen to EVs in charging stations, communication, and battery charging systems.

The CPS can reduce the possibility of occurrence and consequences of cyber-attacks. Ability to monitor the charging stations, integrity of data and information, and confidentiality of transferred data are the main criteria in designing a CPS.

In order to have a better view of security challenges, the concept of cybersecurity analysis has been introduced. The cybersecurity analysis of charging EVs includes the physical and cyber layers. The electrical and communications issues are investigated in the physical and cyber layers, respectively.

Cybersecurity challenges in EV communication include the limited connectivity, limited computational performance, threats of unpredictable attacks, and critical hazard to the life of drivers and passengers. It should be noted that in order to make a better decision, the targets of cyberattacks on EVS should be examined. The control systems, driving systems, and V2X communication can be targeted by cyberattacks.

In order to achieve a secure system against cyberattacks, the following issues are important and should be considered in future studies:

- Cryptographic solutions to perpetuating security services
- Protecting cyber-physical EV charging systems
- Strong authentication
- Reinforcing security policy
- Smart cooperation
- Observing users' privacy
- Prioritizing and categorizing of information

- Dynamic key
- Confirmed and secure backup
- Security check
- Risk measurement
- Artificial intelligent security measures

4.4 Learned Lessons and Concluding Remarks

In order to ensure the security of CPSs, like EV charging stations, the following points should be taken into account [61]:

- Perpetuating security services: to secure the charging process of EVs in a real-time manner, with the least possible amount of computational burden, new lightweight cryptographic solutions are required. Such solutions can help in making sure of the following security services:
  - Using cryptographic algorithms so as to secure real-time communication (confidentiality): Developing dynamic passwords requiring two iterations with only a few operations at a physical layer is an example of these cryptographic algorithms.
  - Protecting cyber-physical EV charging systems and the integrity of equipment against any physical or logical change: Holding the secure design of programs and software in operating systems and the deployment of strong cryptographic functions are examples of such actions.
  - Availability of the information of devices that are engaged in the process of charging vehicles: There are requirements for computational resources, as well as verified backups, and the ability to independently restore the system immediately after cyber- or physical attacks that should be provided.
  - Strong user/device authentication: An effective multifactor user/device authentication along with an increase in the stages of confirmation and identification in accordance to credited access-control entitlements (least entitlement) should be implemented in order to guarantee non-repudiation when an attack occurs.
  - Reinforcing security policy: In most cases, attacks are carried out by insiders (accidentally or intentionally). Therefore, all employees must go through a scanning process prior to being recruited and be denied to have access during outside office hours.
  - Smart cooperation using non-cryptographic methods: Intrusion detection systems should efficiently cooperate with other security solutions, such as firewalls and deceptive techniques like dynamic honeypots.
  - Observing users' privacy: It is essential to protect users' rights over their personal data from violation and strike a balance between users' privacy and the security, as well as the performance of CPSs.

- In order to enhance the level of protection against possible threats and various attacks, one can adopt and enforce such different security strategies as follows:

  - Prioritizing and categorizing: Prior to analyzing, managing, and assessing security risks according to the proportion of their cost to the probability of their occurrence and their impact, it is important for critical components and assets of EV charging systems to be ranked and classified, thus making sure of allocating budget properly on wisely selected security measures/countermeasures.
  - Cryptographic algorithms dependent on dynamic keys: These algorithms can be employed to make sure of not only confidentiality services but also integrity and authentication services. As there is one dynamic key for each message (or a message set; depending upon the limitations and conditions of an application), these algorithms are proved resilient to the occurrence and propagation of attacks.
  - Access control through defining entitlements: Access privilege should be granted in proportion to the role, duties, and qualities of a user when he/she accesses an EV charging system.
  - Strong multifactor schemes for the authentication of entities: It is unfortunate that single-factor entity authentication schemes are not robust enough to resist rapidly increasing authentication attacks. Authentication schemes, the first defense line, are at the forefront of every system. Hence, any authentication attack can lead to a confidentiality attack, integrity attack, and/or availability attack. Currently, a multifactor authentication concept that takes advantage of the combination of two or more factors has been employed.
  - Strong password and dynamic hashing process: In an attempt to make sure of the prevention of some attacks, the regular-based rehashing of passwords with the new one is crucial for every user.
  - Protected and secure audit: This examination can be performed through an audit-managing system that gathers and archives logs in a distributed system. Such solutions not only restrain insiders' ability to conduct an attack on CPSs but also store the electronic testimony of external and internal attacks to find their origins later.
  - Confirmed and secure backup: Making a copy of an EV charging system's information is essential for preserving the availability of data, preventing data from damage and manipulation, and ensuring resiliency against DoS attacks.
  - Security check: In order to prevent the possibility of an insider attack, every employee should be inspected and scanned before and during the office time. Therefore, it is strongly recommended that employees sign a confidentiality agreement, nondisclosure agreement, confidential disclosure agreement, secrecy agreement, or proprietary information agreement. These measures, especially in crucial sections, are of vital importance.
  - Periodic risk measurement: To be capable of classifying the risk of a CPS, like EVSC, according to permissible/non-permissible level and to minimize

the risk as immediately as possible, a regular risk assessment should be performed. The evaluation of the probability and level of a given risk against the system can be conducted by such methods as quantitative and/or qualitative risk measurements and benefit to cost analysis.

- Up-to-date systems: It is crucial for a CPS, like EVSCs, to be integrated into a smart power system and updated in terms of software, hardware, and firmware. Additionally, these systems should be protected at different implementation layers and should be capable of mitigating and repelling a given attack, minimizing the impact of an attack, de-escalating it, and avoiding further destruction. Additionally, USB ports should be removed physically and logically. Moreover, it is important to constantly control how PLC systems behave and perform so as to detect any skeptical and unusual behavior.
- Artificial intelligent security measures: Abnormality identification schemes could be done by taking the advantages of artificial intelligence (AI). AI offers a higher level of accuracy in the detection and prevention of attacks compared to the human intervention methods. Current progress in machine learning, particularly in deep learning, can pave the way for strongly immunizing EV charging systems integrated into the power systems against any cyber-physical attacks.

References

1. Das, H. S., Tan, C. W., & Yatim, A. (2017). Fuel cell hybrid electric vehicles: A review on power conditioning units and topologies. *Renewable and Sustainable Energy Reviews, 76*, 268–291.
2. Mi, C., & Masrur, M. A. (2017). *Hybrid electric vehicles: Principles and applications with practical perspectives*. Wiley.
3. Hannan, M., Azidin, F., & Mohamed, A. (2014). Hybrid electric vehicles and their challenges: A review. *Renewable and Sustainable Energy Reviews, 29*, 135–150.
4. Thomas, C. (2009). Fuel cell and battery electric vehicles compared. *International Journal of Hydrogen Energy, 34*, 6005–6020.
5. Kempton, W. (2016). Electric vehicles: Driving range. *Nature Energy, 1*, 1–2.
6. Wu, H. H., Gilchrist, A., Sealy, K. D., & Bronson, D. (2012). A high efficiency 5 kW inductive charger for EVs using dual side control. *IEEE Transactions on Industrial Informatics, 8*, 585–595.
7. Zhang, L., Brown, T., & Samuelsen, S. (2013). Evaluation of charging infrastructure requirements and operating costs for plug-in electric vehicles. *Journal of Power Sources, 240*, 515–524.
8. Zhang, C., Huang, Q., Tian, J., Chen, L., Cao, Y., & Zhang, R. (2011). Smart grid facing the new challenge: The management of electric vehicle charging loads. *Energy Procedia, 12*, 98–103.
9. Aghapour, R., Sepasian, M. S., Arasteh, H., Vahidinasab, V., & Catalão, J. P. (2020). Probabilistic planning of electric vehicles charging stations in an integrated electricity-transport system. *Electric Power Systems Research, 189*, 106698.

10. Lin, C.-C., Deng, D.-J., Kuo, C.-C., & Liang, Y.-L. (2017). Optimal charging control of energy storage and electric vehicle of an individual in the internet of energy with energy trading. *IEEE Transactions on Industrial Informatics, 14*, 2570–2578.
11. Guo, L., Yang, B., Ye, J., Chen, H., Li, F., Song, W., et al. (2020). Systematic assessment of cyber-physical security of energy management system for connected and automated electric vehicles. *IEEE Transactions on Industrial Informatics, 17*, 3335–3347.
12. Mousavian, S., Erol-Kantarci, M., & Mouftah, H. T. (2018). Cyber-security and resiliency of transportation and power systems in smart cities. In *Transportation and power grid in smart cities: Communication networks and services* (pp. 507–527).
13. Chandwani, A., Dey, S., & Mallik, A. (2020). Cybersecurity of onboard charging systems for electric vehicles—Review, challenges and countermeasures. *IEEE Access, 8*, 226982–226998.
14. E. U. A. f. N. a. I. Security. (2013, December 9). *Smart grid threat landscape and good practice guide*.
15. Dagoumas, A. (2019). Assessing the impact of cybersecurity attacks on power systems. *Energies, 12*, 725.
16. Wise, D. (2016). Vehicle cybersecurity: DOT and industry have efforts under way but DOT needs to define its role in responding to a real-world attack. *Gao Reports. US Government Accountability Office*.
17. El-Gleel, M. A., El-Zawawi, A. M. O., & El-Habrouk, M. Secure lightweight CAN protocol handling message loss for electric vehicles. In *2021 International Conference on Communication, Control and Information Sciences (ICCISc)* (Vol. 2021, pp. 1–6).
18. Basnet, M., & Ali, M. H. (2020). Deep learning-based intrusion detection system for electric vehicle charging station. In *2020 2nd International Conference on Smart Power & Internet Energy Systems (SPIES)* (pp. 408–413).
19. Basnet, M., & Ali, M. H. (2021). Exploring cybersecurity issues in 5G enabled electric vehicle charging station with deep learning. *arXiv preprint arXiv:2104.08553*.
20. ElHusseini, H., Assi, C., Moussa, B., Attallah, R., & Ghrayeb, A. (2020). Blockchain, AI and smart grids: The three musketeers to a decentralized EV charging infrastructure. *IEEE Internet of Things Magazine, 3*, 24–29.
21. Su, Z., Wang, Y., Xu, Q., Fei, M., Tian, Y.-C., & Zhang, N. (2018). A secure charging scheme for electric vehicles with smart communities in energy blockchain. *IEEE Internet of Things Journal, 6*, 4601–4613.
22. *Top Five Blockchain Security Issues in 2019*. Available: https://ledgerops.com/blog/2019-03-28-top-five-blockchain-security-issues-in-2019/
23. Liu, M., Teng, Y., Yu, F. R., Leung, V. C., & Song, M. (2019). Deep reinforcement learning based performance optimization in blockchain-enabled internet of vehicle. In *ICC 2019–2019 IEEE International Conference on Communications (ICC)* (pp. 1–6).
24. Acharya, S., Dvorkin, Y., & Karri, R. (2020). Public plug-in electric vehicles+ grid data: Is a new cyberattack vector viable? *IEEE Transactions on Smart Grid, 11*, 5099–5113.
25. Acharya, S., Dvorkin, Y., & Karri, R. (2021). Causative cyberattacks on online learning-based automated demand response systems. *IEEE Transactions on Smart Grid*.
26. Khan, O. G. M., El-Saadany, E., Youssef, A., & Shaaban, M. (2019). Impact of electric vehicles botnets on the power grid. In *2019 IEEE Electrical Power and Energy Conference (EPEC)* (pp. 1–5).
27. Morrison, G. S. (2018). *Threats and mitigation of DDoS cyberattacks against the US power grid via EV charging*. Wright State University.
28. Gottumukkala, R., Merchant, R., Tauzin, A., Leon, K., Roche, A., & Darby, P. Cyber-physical system security of vehicle charging stations. In *2019 IEEE Green Technologies Conference (GreenTech)* (Vol. 2019, pp. 1–5).
29. Chen, L., Zhou, J., Chen, Y., Cao, Z., Dong, X., & Choo, K.-K. R. (2020). PADP: Efficient privacy-preserving data aggregation and dynamic pricing for vehicle-to-grid networks. *IEEE Internet of Things Journal, 8*, 7863–7873.
30. Olufowobi, H., & Bloom, G. (2019). Connected cars: Automotive cybersecurity and privacy for smart cities. In *Smart cities cybersecurity and privacy* (pp. 227–240). Elsevier.

31. Dey, S., & Khanra, M. (2020). Cybersecurity of plug-in electric vehicles: Cyberattack detection during charging. *IEEE Transactions on Industrial Electronics, 68*, 478–487.
32. Learned, L. (2019). *Risks posed by firewall firmware vulnerabilities*. Tech. rep. North American Electric Reliability Corporation.
33. Sookhak, M., Tang, H., He, Y., & Yu, F. R. (2018). Security and privacy of smart cities: A survey, research issues and challenges. *IEEE Communications Surveys & Tutorials, 21*, 1718–1743.
34. Cui, L., Xie, G., Qu, Y., Gao, L., & Yang, Y. (2018). Security and privacy in smart cities: Challenges and opportunities. *IEEE Access, 6*, 46134–46145.
35. Acharya, S., Dvorkin, Y., Pandžić, H., & Karri, R. (2020). Cybersecurity of smart electric vehicle charging: A power grid perspective. *IEEE Access, 8*, 214434–214453.
36. Denton, T. (2020). *Electric and hybrid vehicles*. Routledge.
37. Emadi, A. (2014). *Advanced electric drive vehicles*. CRC Press.
38. Kim, K., Kim, J. S., Jeong, S., Park, J.-H., & Kim, H. K. (2021). Cybersecurity for autonomous vehicles: Review of attacks and defense. *Computers & Security*, 102150.
39. Koscher, K., Savage, S., Roesner, F., Patel, S., Kohno, T., Czeskis, A., et al. Experimental security analysis of a modern automobile. In *2010 IEEE Symposium on Security and Privacy* (Vol. 2010, pp. 447–462).
40. Halahan, J., & Chen, W. (2017). Wireless security within new model vehicles. *Journal of Information Warfare, 16*, 51–62.
41. Hoppe, T., & Dittman, J. (2007). Sniffing/Replay Attacks on CAN Buses: A simulated attack on the electric window lift classified using an adapted CERT taxonomy. In *Proceedings of the 2nd Workshop on Embedded Systems Security (WESS)* (pp. 1–6).
42. Pan, L., Zheng, X., Chen, H., Luan, T., Bootwala, H., & Batten, L. (2017). Cyber security attacks to modern vehicular systems. *Journal of Information Security and Applications, 36*, 90–100.
43. Wyglinski, A. M., Huang, X., Padir, T., Lai, L., Eisenbarth, T. R., & Venkatasubramanian, K. (2013). Security of autonomous systems employing embedded computing and sensors. *IEEE Micro, 33*, 80–86.
44. Malhi, A. K., Batra, S., & Pannu, H. S. (2020). Security of vehicular ad-hoc networks: A comprehensive survey. *Computers & Security, 89*, 101664.
45. Saadat, S., Maingot, S., & Bahizad, S. (2020). Electric vehicle charging station security enhancement measures. In *2020 5th IEEE Workshop on the Electronic Grid (eGRID)* (pp. 1–8).
46. Harnett, K., Harris, B., Chin, D., & Watson, G. (2018). *Doe/dhs/dot volpe technical meeting on electric vehicle and charging station cybersecurity report*. John A. Volpe National Transportation Systems Center (US).
47. Bhaskar, N., Bland, M., Levchenko, K., & Schulman, A. (2019). Please pay inside: Evaluating bluetooth-based detection of gas pump skimmers. In *28th {USENIX} Security Symposium ({USENIX} Security 19)* (pp. 373–388).
48. Van Aubel, P., Poll, E., & Rijneveld, J. (2019). Non-repudiation and End-to-End security for electric-vehicle charging. In *2019 IEEE PES Innovative Smart Grid Technologies Europe (ISGT-Europe)* (pp. 1–5).
49. Alcaraz, C., Lopez, J., & Wolthusen, S. (2017). OCPP protocol: Security threats and challenges. *IEEE Transactions on Smart Grid, 8*, 2452–2459.
50. Xiao, Q., Gibbons, T., & Lebrun, H. (2009). Rfid technology, security vulnerabilities, and countermeasures. In *Supply chain the way to flat organization* (pp. 357–382). Publisher-Intech.
51. El-Rewini, Z., Sadatsharan, K., Selvaraj, D. F., Plathottam, S. J., & Ranganathan, P. (2020). Cybersecurity challenges in vehicular communications. *Vehicular Communications, 23*, 100214.
52. Danish, S. M., Zhang, K., Jacobsen, H.-A., Ashraf, N., & Qureshi, H. K. (2020). BlockEV: Efficient and secure charging station selection for electric vehicles. *IEEE Transactions on Intelligent Transportation Systems*.
53. Bao, K., Valev, H., Wagner, M., & Schmeck, H. (2018). A threat analysis of the vehicle-to-grid charging protocol ISO 15118. *Computer Science-Research and Development, 33*, 3–12.

54. Yang, B., Guo, L., Li, F., Ye, J., & Song, W. Impact analysis of data integrity attacks on power electronics and electric drives. In *2019 IEEE Transportation Electrification Conference and Expo (ITEC)* (Vol. 2019, pp. 1–6).
55. Axell, E., Eliardsson, P., Tengstrand, S. Ö., & Wiklundh, K. (2016). Power control in interference channels with class a impulse noise. *IEEE Wireless Communications Letters, 6*, 102–105.
56. Reeh, D., Tapia, F. C., Chung, Y.-W., Khaki, B., Chu, C., & Gadh, R. Vulnerability analysis and risk assessment of EV charging system under cyber-physical threats. In *2019 IEEE Transportation Electrification Conference and Expo (ITEC)* (Vol. 2019, pp. 1–6).
57. Chan, A. C.-F., & Zhou, J. (2015). A secure, intelligent electric vehicle ecosystem for safe integration with the smart grid. *IEEE Transactions on Intelligent Transportation Systems, 16*, 3367–3376.
58. Mousavian, S., Erol-Kantarci, M., & Ortmeyer, T. Cyber attack protection for a resilient electric vehicle infrastructure. In *2015 IEEE Globecom Workshops (GC Wkshps)* (Vol. 2015, pp. 1–6).
59. Lee, A. (2010). *Guidelines for smart grid cyber security.*
60. Egozcue, E., Rodriguez, D., Ortiz, J., Villar, V., & Tarrafeta, L. (2012). *Smart grid security—Recommendation for Europe and member states, ENISA.*
61. Yaacoub, J.-P. A., Salman, O., Noura, H. N., Kaaniche, N., Chehab, A., & Malli, M. (2020). Cyber-physical systems security: Limitations, issues and future trends. *Microprocessors and Microsystems, 77*, 103201.

Chapter 5
Evaluation of Cyberattacks in Distribution Network with Electric Vehicle Charging Infrastructure

Uldouz Bazazi and Sajad Najafi Ravadanegh

5.1 Introduction

A smart grid is a power grid that utilizes two-way information exchange and secure telecommunication technology at all levels of power systems, including production, transmission, distribution, and consumers to achieve a secure, clean, reliable, efficient, sustainable, and resilient network. The resilience and security of these networks are always at risk by both natural disasters such as earthquakes, storms, and tornadoes and intentional factors such as cyberattacks. Cyberattacks are a type of attack that aims to destroy or manipulate information and data related to smart grids. Lack of protection against these attacks can lead to disruption of the performance of various parts of the network. The intrusion of these attacks can start from different infrastructures according to their communication networks, and the infrastructures of EVSEs are vulnerable parts that can be the point of penetration of the attackers. These attacks, which spread contamination through EVSEs and EVs, can have unforeseen impacts for EVs, as well as more severe effects such as power outages. Therefore, to maintain the security and efficiency of the network, these cyberattacks and infrastructure must be properly modeled, and a protection plan must be provided to deal with them. The security and resiliency of smart grids are threatened in the presence of electric vehicle infrastructure which includes EVs, EVSEs, meters, and other roadside infrastructures that use telecommunication technologies to exchange information, which are vulnerable to cyberattacks. The mobility and random movement of the EVs in EVSE infrastructure and the application of wireless communication network for charging platforms

U. Bazazi · S. N. Ravadanegh (✉)
Resilient Smart Grids Research Lab, Electrical Engineering Department, Azarbaijan Shahid Madani University, Tabriz, Iran
e-mail: s.najafi@azaruniv.edu

V. Vahidinasab, B. Mohammadi-Ivatloo (eds.), *Electric Vehicle Integration via Smart Charging*, Green Energy and Technology,
https://doi.org/10.1007/978-3-031-05909-4_5

can increase the propagation of the attack to the network and vulnerability of the EVSE communication network. Therefore, due to the increase in the number of EVs in the future, the penetration through this infrastructure should be prevented, and the necessary measures should be taken to accurately identify the infrastructure and cyberattacks and deal with the threats related to these attacks.

5.2 Status Quo, Challenges, and Outlook

Electric vehicles (EVs) are considered as the mobile portion of the smart grid. The energy stored in the batteries of EVs, in addition to providing their propulsion, can have other applications such as ancillary services including peak power supply and short-term or emergency power supply in homes and businesses due to the use of smart grid communication technologies [1]. EVs, electric vehicle supply equipment (EVSEs), meters, and other roadside equipment form electric vehicle infrastructure (EVI). EVSEs are the connection points of EVs to the power grid [2]. Various communication technologies like Wi-Fi, Zigbee, 3G/4G, and dedicated short-range communication (DSRC) are used in different communication parts of EVI including charging EVs in EVSEs, EVSEs shared communication network, and EVs shared communication network [3, 4]. Telecommunication infrastructure and communication technologies are used to swapping basic information such as state of charge (SoC), optimal charging time, energy price, load control signals from EVSEs to EVs, and the distance from EV to EVSE [5, 6]. Also, EVSEs share usage data to improve the operator's services and functions [2]. EVs share traffic status, warning signals, road safety for traffic improvement, and reducing the risk of accidents [7]. The general EVSEs shared communication network is illustrated in Fig. 5.1.

Despite many benefits and opportunities created by EVI, if their security gaps are not properly addressed, they can pose significant threats to the smart grid and

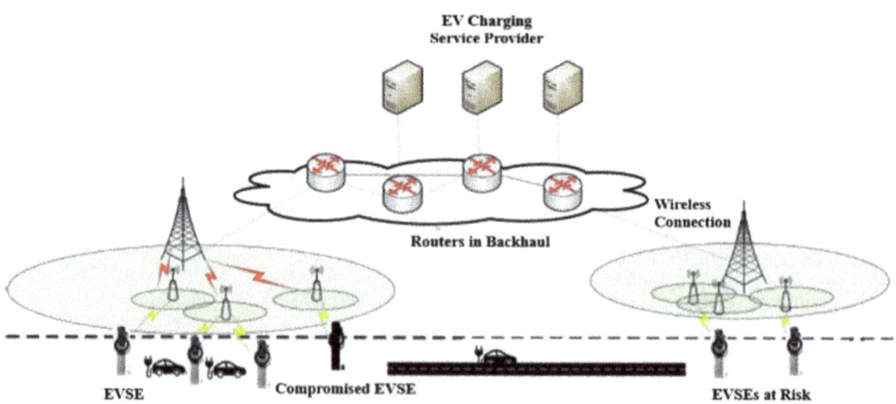

Fig. 5.1 EVSEs shared communication network

jeopardize its resilience. Physical manipulation, malicious software uploads, and load changes are among the possible cyber-physical attacks on EVI [6, 8]. This malware depending on the attacker's aim and malicious plan may not be limited to this infrastructure. An ordinary attacker may embed malicious software that runs in the background and slows down the operation of utility systems, whereas at higher levels the attack can infect intelligent equipment such as PMUs, line monitoring, and protection equipment or more severe control generators and by randomly shutting them off cause regional blackouts [2]. Notable point is that the ubiquity of communication networks, as well as the mobility of EVs, accelerates the propagation of cyberattacks initiated by EVIs. So, it would be better to consider the worst-case situation. Among the vulnerabilities that attackers could penetrate EVI, the ECU section in EV, the USB port for updating the EVSE's operating system, the OCPP protocol, and phone applications such as Tesla can be mentioned [9–12]. However, some actions can be taken to reduce cyber threats. The most important of which is the implementation of the OCPP v2.0 protocol and the use of encryption and authenticated control messages in updates.

As regards that smart grids are at risk of various cyberattacks such as traffic analysis, IP scan, viruses and worms, man-in-the-middle (MITM), denial of service (DoS), Trojan horse, backdoor attacks, etc., some articles have provided detection and countermeasure methods for some of these attacks like false data injection (FDI) [13, 14]. Albeit these solutions help in enhancing the smart grid's security, they are inadequate to deal with proficient and blended attacks such as Stuxnet or Duqu. Stuxnet showed that strategies like "defense-in-depth" or "security by obscurity" are no longer credible [15]. So, security is not only achieved through one particular method but also by combining several methods and integrating them. In [14], a cybersecurity strategy consisting of three phases, pre-attack, under-attack, and post-attack, has been proposed. This paper is studied in the under-attack phase that is divided into two tasks including attack detection and attack mitigation. In the attack detection section, all deployed security techniques including IDS, SIMS, and DLP are recommended. Right away after the attack is detected, attack mitigation can be done with several methods especially pushback and reconfiguration [16].

According to the security topics reviewed, the penetration of cyberattacks, and different communication parts of EVI, the spread of an attack can be examined in three ways. In this paper a mixed-integer linear programming (MILP)-based protection and response model is provided to reduce the spread of malware on this infrastructure. As shown in Fig. 5.2, first the contamination levels for EVs and EVSEs are calculated in two intervals, respectively, before identification and before the end of the inspection. Due to the numbers obtained and considering some constraints, to maintain power network availability as well as desirable service, some EVs and EVSEs are temporarily disconnected for further inspection and recovery operations.

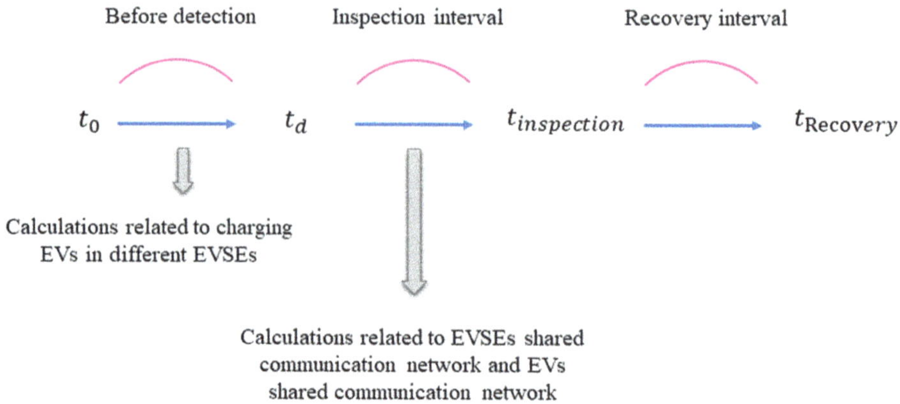

Fig. 5.2 Overview of time intervals

5.2.1 *EV2EVSE*

In the EV charging section, the attack starts from an EV and spreads to other EVs and EVSE during the recharging process. Examples of this attack include software patches from an invalid source or an infected mobile application that is associated with EV and downloads malware [2]. It is assumed that an EVSE is detected infected at t_d. Hence, all EVs connected to the detected EVSE during time $[t_0, t_d]$ as well as EVSEs that are connected for recharging are likely to be compromised. So, the likelihood of being compromised called threat level is calculated in this interval based on the connection pattern. Then, some EVs and EVSEs will be driven out of service provisionally due to the constraints like forecasted demand to maintain optimal service and mitigate the propagation of attack until the end of the inspection [6].

5.2.2 *EVSE2EVSE*

In addition to the EV charging in the various EVSEs, malware or cyberattack can also be spread on the EVSEs shared communication network. It should be noted that in a complex and high-level attack, according to the attacker's goal, infected EVSE or EVSEs can function as an operator and send software updates that are indeed malware to various parts of EVI or even the smart grid. In the previous section, although the EVSE detected by the security alarm and some compromised EVSEs and EVs are detached, other EVSEs can still be compromised due to interplays between EVs and EVSEs until the end of recovery. Thus, recently compromised but undetected EVSEs will carry on to infect more EVSEs through their shared communication network, and spreading the attack on more EVSEs could imperil the infrastructure and power supply of EVs.

5.2.3 EV2EV

The third way that the cyberattack can propagate through is EVs shared communication network. This paper proposes a MILP method that jointly addresses security risk and equipment availability on this network. Communication between EVs is feasible through vehicular ad hoc networks (VANETs). VANETs can use any wireless technology including Wi-Fi, Zigbee, and even cellular technologies. The main standard used is DSRC. DSRC is an open-source protocol for wireless communication that is similar to Wi-Fi in some ways. Wi-Fi is mostly used for wireless local area networks, while DSRC is used for safe and high-speed wireless communication between vehicles, which features a low latency, low interference, good performance in adverse weather conditions, and so on. In fact, the DSRC combines with GPS technology to create a low-cost V2V connection, providing a 360-degree consciousness of the same equipped vehicles within a range of approximately 300 meters. Transmitted messages common to all vehicles include current GPS position; vehicle speed acceleration and heading; vehicle control information such as transmission state, brake status, and steering wheel angle; as well as the vehicle's path history and path prediction [7–10]. A view of common messages between vehicles is shown in Fig. 5.3.

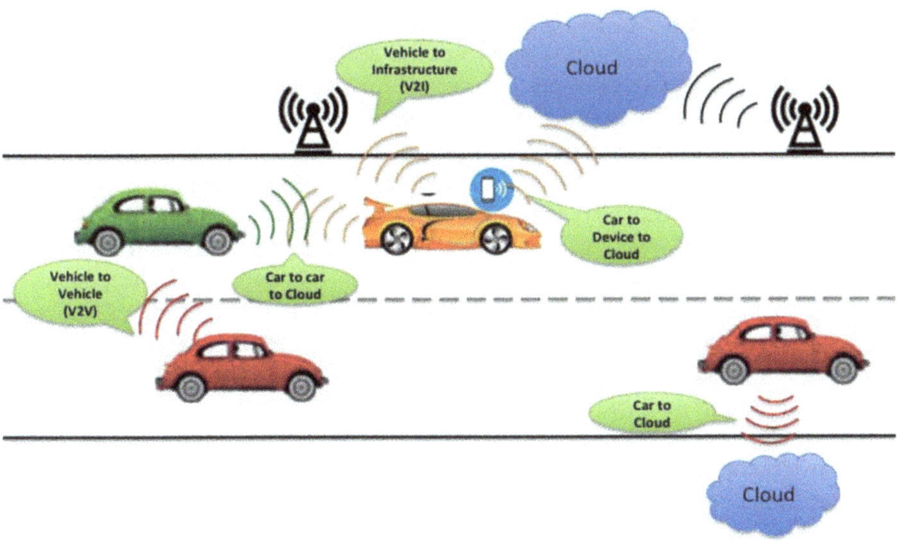

Fig. 5.3 View of common messages between vehicles [8]

The significant matter here is that the mobility model of EVs plays a key role in the propagation of the attack. The model may be such that the more infected EVs are closer to each other and, depending on their network, infection intensifies and the cyberattack spreads more rapidly or vice versa. Like EVSEs communication network, considering some compromised EVs that are isolated after detection, the new threat levels for remaining EVs are computed based on hop distances in the communication network during the time $[t_d, t_{\text{inspection}}]$. The hop distances in the EV network are different at all times due to the varying mobility behavior of EVs. Therefore, we minimize the threat levels by detaching another subset of EVs to lessen the venture of the attack until the end of the recovery.

5.3 Related Work

As mentioned in the previous parts, the smart power grid has many electrical components and equipment, as well as communication and automation equipment, including electrical relays, monitoring, and control devices. Recent Risk Management Process (RMP) guidelines developed by the Department of Energy (DOE), National Institute of Standards and Technology (NIST), and North American Electric Reliability Corporation (NERC) state that common border conservation techniques, accompanying of information and communication technologies (ICT), and increasing the permeability of cyberattacks are no longer effective in the energy sector [17]. This is because the attackers can interpenetrate any cyber-physical energy infrastructure and lead the whole grid to a fickle condition. Ensuring that all equipment is safe, despite having security measures to prevent attacks, is very challenging. Adding EVI to this interconnected space of equipment obliges the smart grid to have preparedness and response strategies to reduce the impacts of these attacks or at least their rate of spread. Despite the many articles that have been written on the smart grid and EVI safety, few studies have been done on the security vulnerabilities that result from their affiliation. In [18], the EV passes the challenge-response verification utilizing the trusted computing base (TCB). A malicious load is physically connected to the grid instead of EV and does not react to the demand-response commands, and simultaneous operation of multiple loads triggers the circuit breakers. Therefore, the incongruity of the physically connected device and the cyber-certified EV jeopardize power grid resiliency. In [19], the attack on the dynamic price of electricity is considered and its effects on EV charging are discussed. Price information attacks affect the optimization of EV charging programs and lead to differences between the actual and expected costs of EV charging. This problem is raveled using a linear programming (LP) technique to achieve an optimal charging policy against these attacks. In [20], the impacts of malware infection on the upmost equipment of the smart grid, including EVSEs, utility systems, transformers, circuit breakers, etc., have been studied. Security certificates are proposed as a solution but can be stolen through malware so that malicious users can authenticate malware as licit software and endanger the system.

In [21], a specific type of MITM cyber-physical attack on the smart grid and EVI has been addressed, and a two-step challenge-response plan has been proposed to mitigate it. In [22], the first is focused on typical cybersecurity issues as well as smart grid integration with EVs. Then it introduces a new class of key settlement protocols and a jamming-resilient protocol for wireless networks. The key protocols do not need security substructure support due to their capability of generating inexpensive and random secret keys. In [23], an optimal response model based on MILP is proposed against cyberattacks on phasor measurement units (PMU) [24]. Uses a novel protection method by the junction of attack analysis and contingency analysis. Types of probabilistic response models have been discussed in [6–23, 25]. Also, load alteration and bad data injection attacks have been studied in [26, 27]. In [28], the authors have formulated an optimized MILP-based model to eschew propagation of the cyberattacks in an open-science computer network in which network sites have communication with each other. This model defines the sites that should be interrupted from the network to increase the number of connected users due to the certain threshold value. In [2–6], the basic forms of this paper, namely, an attack model that presumes spread through charging and EVSEs shared communication network, are proposed. In this paper, we develop the model to include attack spread through EVs shared communication network.

5.4 Cyberattack Model

Among the usual v2v attacks, we can name DoS and FDI [7]. As mentioned earlier, EVSEs detected by security alarms, as well as some EVSEs and EVs with the determined threat levels at t_d and according to the relevant constraints, were isolated temporarily to mitigate the process of spreading the attack until the end of the inspection. However the infection level of remaining EVs whether compromised or uncompromised such as EVSEs can still increase due to the interaction with EVSEs in the charging process. Some of the cyberattacks like worms and viruses do not operate until they are fully spread so that can be another cause of being tardy in the detection of malware propagation and therefore unrecognizable. Therefore, these EVs that have not yet been detected will continue to infect more EVs through their shared communication network until the end of recovery. The spread of the attack on more EVs, depending on the complexity and severity of the attack, could endanger the EVI, power supply, and even the security and availability of the smart grid. So, since the results of the attack may not be recognizable at the time of detection, it would be reasonable to assume the worst-case scenario that the attacker targeted the most severe damages. We first formulate the attack propagation, to prevent such situations. The summary of used notations can be seen in the nomenclature at the beginning of the chapter.

We consider that some EVs with the likelihood of being compromised have been selected for taking out of service temporarily after detection time t_d based on the response model implementation in EV2EVSE. On the other hand, the remaining

EVs are likely to be infected but not detected during the time $[t_d, t_{\text{inspection}}]$. The likelihood of an EV being compromised or threat level at the time t has been shown by the $\beta_j(t)$. We need $\beta_j(t_d)$ as initial threat levels of EVs that have been calculated in the EV2EVSE section for calculations at times after t_d.

As explained before, after detection, some EVs are isolated, but the attack keeps on to outbreak through the remaining EVs' shared communication network. We represent the time that an attack spread takes in the EV communication network by Δt and the inspection interval by $K\Delta t$ [2]. Equation (5.1) shows the estimation of threat levels of EVs at time $t_d + \Delta t$:

$$\begin{aligned} \beta_j(t_d + \Delta t) &= \Pr\left(\text{EM}_j(t_d + \Delta t) = 1\right) \\ (B) \quad &= \Pr\left\{\text{EM}_j(t_d + \Delta t) = 1 | \text{EM}_j(t_d) = 0\right\} \\ \text{(i)} \quad &\times \Pr\left\{\text{EM}_j(t_d) = 0\right\} \\ \text{(ii)} \quad &+ \Pr\left\{\text{EM}_j(t_d + \Delta t) = 1 | \text{EM}_j(t_d) = 1\right\} \\ \text{(iii)} \quad &\times \Pr\left\{\text{EM}_j(t_d) = 1\right\} \quad \forall j \in \text{X} \end{aligned} \tag{5.1}$$

In which $\beta_j(t)$ is the probability of an EV being compromised, Δt is the time that an attack propagated in the EV communication network, and t_d is the time of detection. $\text{EM}_j(t)$ is a random variable which equals 1 if EV_j is compromised at time t, and 0 otherwise. U_j is a random variable which equals 1 if a cyberattack propagates to and compromises EV_j, and 0 otherwise. U_{ijk} is a random variable which equals to 1 if a cyberattack propagates from EV_i and targeting EV_j compromises the kth communication relay between EV_i and EV_j, and 0 otherwise. $K\Delta t$ is the inspection time interval and Pr is the probability function.

The last term of the right-hand side, (iii), equals to $\beta_j(t_d)$, so the second term (i) equals to $1 - \beta_j(t_d)$ and the third term (ii) equals to 1. The first term (B) shows the probability of being compromised at $t = t_d + \Delta t$, given being uncompromised at $t = t_d$, as follows:

$$\begin{aligned} B &= 1 - \Pr\left\{\text{EM}_j(t_d + \Delta t) = 0 | \text{EM}_j(t_d) = 0\right\} \\ &= 1 - \prod_{\substack{i \in \text{X} \\ i \neq j}} \left(\Pr\left\{\text{EM}_j(t_d + \Delta t) = 0 | \text{EM}_i(t_d) = 1\right\}\right) \\ &= 1 - \prod_{\substack{i \in \text{X} \\ i \neq j}} \left(1 - \zeta_{ij}\beta_i(t_d)\right) \end{aligned} \tag{5.2}$$

In (5.2), we used complementary probabilities. In other words, to get B, we need to compute $\Pr\{\text{EM}_j(t_d + \Delta t) = 1 | \text{EM}_i(t_d) = 1\}$, which means the probability that EV_j becomes infected given that EV_i is infected. So, according to the statistical dependency of these random variables ($\forall i \neq j$), complementary probabilities are used, and the term $\Pr\{\text{EM}_j(t_d + \Delta t) = 0 | \text{EM}_i(t_d) = 1\}$ is computed instead due to their statistical independence $i \neq j$. This term shows the probability of EV_j does not become infected given that EV_i is infected. In the last term, ζ_{ij} is the probability that the attack spreads from infected EV_i to uninfected EV_j during the Δt as shown in Eq. (5.3):

$$\begin{aligned}\zeta_{ij} &= \Pr\left(U_j = 1\right) \times \prod_{k=1}^{\delta_{ij}} \Pr\left\{U_{ijk} = 1\right\} \\ &= \lambda \times \prod_{k=1}^{\delta_{ij}} \varphi \\ &= \lambda \varphi^{\delta_{ij}} \; i, j \in \mathrm{X}\end{aligned} \tag{5.3}$$

In Eq. (5.3), λ or $\Pr(U_j = 1)$ indicates the probability of EV compromisation by a transmitted attack at the destination. φ or $\Pr\{U_{ijk} = 1\}$ is the probability that an attack spreads toward the purposed EV through a communication relay in the EV communication network. δ_{ij} represents the hop distance, and it is the minimum number of communication relays that could be a router and link EV_i and EV_j. Unlike EVSEs, which have a fixed location, EVs are moving at all times and have different spatial coordinates; so the hop distances between them and subsequently δ_{ij} are also different due to their mobility model at all times. The shorter the hop distance, the faster the attack spread. In other words, the threat level of an EV enhances in more rapid steps because of its short hop distance to other EVs although its initial threat level may be small. By placing Eqs. (5.2) and (5.3) in Eq. (5.1) and simplifying the expressions, Eq. (5.4) is obtained which presents the threat level of EV_j at the time$t = t_d + \Delta t$:

$$\beta_j\left(t_d + \Delta t\right) = 1 - \left(1 - \beta_j\left(t_d\right)\right) \prod_{\substack{i \in \mathrm{X} \\ i \neq j}} \left(1 - \zeta_{ij}\beta_i\left(t_d\right)\right), \forall j \in \mathrm{X} \tag{5.4}$$

The threat levels of EVs increase during the time until the network fully recovers. The general threat level formula considering the initial threat level and Eq. (5.4) is obtained for a period of time in Eq. (5.5). This period can be the inspection interval. We will use the attack model and threat levels in the response model:

$$\begin{aligned}\beta_j\left(\text{initial}\right) &= \beta_j\left(t_d\right) \\ \beta_j\left(t_d + n\Delta t\right) &= 1 - \left(1 - \beta_j\left(t_d + (n-1)\Delta t\right)\right) \\ &\times \left(\prod_{\substack{i \in \mathrm{X} \\ i \neq j}} \left(1 - \beta_i\left(t_d + (n-1)\Delta t\right) \times \zeta_{ij}\right) \right) \\ &\forall j \in \mathrm{X} \; ; \; 1 \leq n \leq K\end{aligned} \tag{5.5}$$

5.4.1 Response Model

The common action taken when a cyberattack occurs in infrastructure is to isolate elements detected by the detection alarm. Our proposed model represents that the compromised ones as well as those detected are at high risk of being infected and should be taken out of service temporarily due to security considerations. So, in the EV network as well, the goal is to find the EVs that are most likely to be infected and assist the attack spread as much as possible but have not yet been detected. By isolating these EVs temporarily until the network recovery is complete, the pace and rate of attack propagation can be reduced to some extent.

The goal of the inspection period is that after the detection alarm shuts down, the section operator requires a specific time span for checking and investigating the network, during which time the attack continues to spread. Therefore, according to the relevant constraints, some EVs with the highest probability of compromising at the time $t = t_d$ were disconnected by the time of inspection, which is given in EV2EVSE.

Upon the end of the inspection, i.e., at $t_d + K\Delta t$ or $t_{\text{inspection}}$, EVs that still existed and are potentially compromised will keep on to propagate the attack and infection until appropriate security patches are uploaded and set up from a trusted source, usually remotely, and the network is fully recovered.

The suggested response scheme is adjusted based on a mixed-integer linear programming (MILP) model that ascertains which EVs should be disconnected, and as an outcome of that, the maximum threat level of remaining EVs is minimized until the end of recovery. A binary decision variable Zev_j is needed in threat levels formulation that can separate the disconnected EVs:

$$\begin{aligned} \beta_j\,(\text{initial}) &= \beta_j\,(t_d) \\ \beta_j\,(t_d + n\Delta t) &= 1 - \left(1 - \beta_j\,(t_d + (n-1)\,\Delta t)\right) \\ &\quad \times \left(\prod_{\substack{i \in \mathrm{X} \\ i \neq j}} \left(1 - \beta_i\,(t_d + (n-1)\,\Delta t) \times \zeta_{ij} \times \text{Zev}_j\right) \right) \\ &\quad \forall j \in \mathrm{X}\ ;\ \ 1 \le n \le K \end{aligned} \tag{5.6}$$

Equation (5.6) is nonlinear and to be able to solve it efficiently, the linear equivalent Eq. (5.7) is obtained as follows:

$$\begin{aligned} \ln\left(1 - \beta_j\,(t_d + n\Delta t)\right) = &\sum_{\substack{i \in \mathrm{X} \\ i \neq j}} \ln\left(1 - \beta_i\,(t_d + (n-1)\,\Delta t)\,\zeta_{ij}\text{Zev}_j\right) \\ &+ \ln\left(1 - \beta_j\,(t_d + (n-1)\,\Delta t)\right) \end{aligned} \tag{5.7}$$

To continue, we use the following equality (5.8) where G is a constant and Zev_j is a binary variable in (5.7), to obtain final Eq. (5.9) [23]:

$$\ln\left(1 - G\mathrm{Zev}_j\right) = \mathrm{Zev}_j \ln\left(1 - G\right) \tag{5.8}$$

$$\begin{aligned}\ln\left(1 - \beta_j\left(t_d + n\Delta t\right)\right) = & \sum_{\substack{i \in \mathrm{X} \\ i \neq j}} \mathrm{Zev}_i \ln\left(1 - \beta_i\left(t_d + (n-1)\Delta t\right)\zeta_{ij}\right) \\ & + \ln\left(1 - \beta_j\left(t_d + (n-1)\Delta t\right)\right)\end{aligned} \tag{5.9}$$

The objective function of the response model is the minimization of the maximum threat level of all connected EVs at the end of inspection $t = t_d + K\Delta t$:

$$\mathrm{P_{EV}} = \min_{\mathrm{Zev}}\max_{j}\left(\beta_j\left(t_d + K\Delta t\right) \times \mathrm{Zev}_j\right) \qquad \forall j \in \mathrm{X} \tag{5.10}$$

As $\beta_j(t_d + K\Delta t)$ and consequently, the objective function is not linear, we utilized its equivalent linear function and reformulated:

$$\mathrm{P_{EV}} = \min_{\mathrm{Zev}}\max_{j}\left(-\ln\left(1 - \beta_j\left(t_d + K\Delta t\right)\right) \times \mathrm{Zev}_j\right) \qquad \forall j \in \mathrm{X} \tag{5.11}$$

We can define $\rho_j = -\ln\left(1 - \beta_j(t_d + K\Delta t)\right) \times \mathrm{Zev}_j$ and $Y = \max_j\left(\rho_j\right)$. So the objective function can be represented by the following set of linear Eqs. (5.12, 5.13, 5.14, 5.15, and 5.16):

$$\mathrm{P_{EV}} = \min \quad Y \tag{5.12}$$

Subject to:

$$\rho_j \leq \mathrm{Zev}_j \qquad \forall j \in \mathrm{X} \tag{5.13}$$

$$\rho_j \leq -\ln\left(1 - \beta_j\left(t_d + K\Delta t\right)\right) \qquad \forall j \in \mathrm{X} \tag{5.14}$$

$$\rho_j \geq 0 \qquad \forall j \in \mathrm{X} \tag{5.15}$$

$$\rho_j \leq Y \qquad \forall j \in X \tag{5.16}$$

The important matter is that the objective function desires to disconnect all EVs which have positive threat levels. Hence, we add (5.17) to keep the EVs with threat levels less than a threshold value T_{Enet} connected to the network. In other words, an EV to continue with effective and proper functionality must have less threat level than the threshold value. This value is affiliated with the rate of risk taking of power system operators:

$$\beta_j \left(t_d + K\Delta t\right) > T_{\text{Enet}} - \text{Zev}_j \qquad \forall j \in X \tag{5.17}$$

To obtain the linear fashion of Eqs. (5.17, 5.18, and 5.19) have used:

$$1 - \beta_j \left(t_d + K\Delta t\right) < 1 - T_{\text{Enet}} + \text{Zev}_j \tag{5.18}$$

$$\ln\left(1 - \beta_j \left(t_d + K\Delta t\right)\right) < \ln\left(1 - T_{\text{Enet}} + \text{Zev}_j\right) \tag{5.19}$$

Then the right-hand side has simplified:

$$\begin{aligned}\ln\left(1 - \beta_j \left(t_d + K\Delta t\right)\right) < \left(1 - \text{Zev}_j\right) \ln\left(1 - T_{\text{Enet}}\right) \\ + \text{Zev}_j \ln\left(2 - T_{\text{Enet}}\right)\end{aligned} \tag{5.20}$$

$$\begin{aligned}&\sum_{\substack{i \in X \\ i \neq j}} \text{Zev}_i \ln\left(1 - \beta_i \left(t_d + (K-1)\Delta t\right) \zeta_{ij}\right) \\ &+ \ln\left(1 - \beta_j \left(t_d + (K-1)\Delta t\right)\right) \\ &\quad < \left(1 - \text{Zev}_j\right) \ln\left(1 - T_{\text{Enet}}\right) + \text{Zev}_j \ln\left(2 - T_{\text{Enet}}\right)\end{aligned} \tag{5.21}$$

5.5 Experimental Results

We try out the performance of our proposed response model on a 7-EVSE and 21-EV test system and describe how the cyberattack propagates and our scheme against it. In this experimental system, we consider that a cyberattack is detected on $t_d = 50$, and we set $\Delta t = 1$ (s), $t_{\text{inspection}} = 180$ (s), and $t_{\text{Recovery}} = 1500$ (s). The other assumptions are $\lambda = 1$, $\varphi = 0.05$, $T_{\text{Enet}} = 0.3$.

In this system, the EVSE-(1) has been detected compromised at $t_d = 50$, and after implementation of response model in the EV charging section to mitigate the attack until $t_{\text{inspection}}$, some EVSEs along with some EVs, namely, EV-(2) and EV-(14), have been temporarily out of service. After inspection, we run our proposed

Table 5.1 Initial threat levels for EVs

J	β_j	J	β_j
1	0.1092	12	0.0504
2	0.7244	13	0.0585
3	0.5571	14	0.8094
4	0.1487	15	0.0537
5	0.4502	16	0.0493
6	0.0278	17	0.6401
7	0.1348	18	0.0276
8	0.0646	19	0.0680
9	0.6928	20	0.1144
10	0.0240	21	0.1096
11	0.1643	22	

response model to mitigate the attack propagation until the end of recovery. We first need the EVs threat level $\beta_j(t_d)$ as the initial threat level to begin calculations that are represented in Table 5.1. These values impact the whole calculations since the probability of being compromised enhances due to EVs mobility and also EV charging in different EVSEs.

Notice that we assume the EVs have different random mobility patterns at all times, so we have shown randomly generated hop distances in the EV network due to their distance at a specific time as $t = 78$ in Table 5.2. The great hop distance among two EVs causes less attack propagation from one to another because of propagation probability. Now, we can calculate the threat levels. Although several EVs are disconnected from the network as said before, the reaming ones are likely to be infected with probabilities given in Table 5.1 and carry on to more infect other EVs through their shared communication network. The assumed area, an example of the random mobility pattern of EVs and their recharging.

Then, by implementing the optimal response model, another number of EVs is disconnected to control and slacken the attack as much as possible until the end of recovery. The disconnected EVs in this interval include {1, 3, 5, 6, 9, 13, 16, 17, and 18}. Eventually, by comparing the threat levels of remaining EVs with no action and with implementing the proposed model, a decrease in propagation proceeding as well as the level of compromising is observed. The set of remaining EVs, some of which are shown in Fig. 5.4, are as follows: {4, 7, 8, 10, 11, 12, 15, 19, 20, 21}.

According to Fig. 5.4 and considering the initial threat levels, the threat levels of EV-(7) and EV-(12) increment more rapidly due to their shorter hop distances over time. Notice that the threat levels still increase after running the response model. However, the reduction in threat levels is remarkable when the proposed optimal solution is run.

Figure 5.5 shows how effective the parameter φ on threat levels is. We have assumed a data sample space including $n - \text{EV} = 21$, $\Delta t = 1$ (s), $t_{\text{inspection}} = 180$ (s), $t_{\text{Recovery}} = 1500$ (s), and $\lambda = 1$. Given that φ is the probability that an attack propagates through a communication relay, we alter its value from 0.01 to 0.1 and select EV-(6) as an example that can be generalized to other remaining EVs also.

Table 5.2 Hop distances in the EV network at $t = 78$

EV	1	2	3	4	5	6	7	8	9	10	11	12	13	14	15	16	17	18	19	20	21
1	–	0	0	0	0	0	0	0	0	0	0	0	0	0	0	0	0	0	0	0	0
2	0	–	0	0	0	0	0	0	0	0	0	0	0	0	0	0	0	0	0	0	0
3	0	0	–	0	0	0	0	0	0	0	0	0	0	0	0	0	0	0	0	0	0
4	0	0	0	–	0	0	0	0	0	0	0	0	0	0	0	0	0	0	0	0	0
5	0	0	0	0	–	0	3	0	0	0	0	0	0	0	0	0	0	0	0	0	0
6	0	0	0	0	0	–	0	0	0	0	0	0	0	0	0	0	3	0	0	0	0
7	0	0	0	0	3	0	–	0	0	0	0	0	0	0	0	0	0	0	0	0	0
8	0	0	0	0	0	0	0	–	0	0	0	0	0	0	0	0	0	0	3	0	0
9	0	0	0	0	0	0	0	0	–	0	0	0	0	0	0	0	2	0	0	0	0
10	0	0	0	0	0	0	0	0	0	–	0	0	0	0	0	0	0	0	0	0	0
11	0	0	0	0	0	0	0	0	0	0	–	0	0	0	0	0	0	0	0	0	0
12	0	0	0	0	0	0	0	0	0	0	0	–	0	0	0	0	0	0	0	0	0
13	0	0	0	0	0	0	0	0	0	0	0	0	–	0	1	2	0	0	0	0	0
14	0	0	0	0	0	0	0	0	0	0	0	0	0	–	0	0	0	0	0	0	0
15	0	0	0	0	0	0	0	0	0	0	0	0	1	0	–	3	0	0	0	0	0
16	0	0	0	0	0	0	0	0	0	0	0	0	2	0	3	–	0	0	0	0	0
17	0	0	0	0	0	0	0	0	2	0	0	0	0	0	0	0	–	0	0	0	0
18	0	0	0	0	0	0	0	0	0	0	0	0	0	0	0	0	0	–	0	0	2
19	0	0	0	0	0	0	0	3	0	0	0	0	0	0	0	0	0	0	–	0	0
20	0	0	0	0	0	0	0	0	0	0	0	0	0	0	0	0	0	0	0	–	0
21	0	0	0	0	0	0	0	0	0	0	0	0	0	0	0	0	0	2	0	0	–

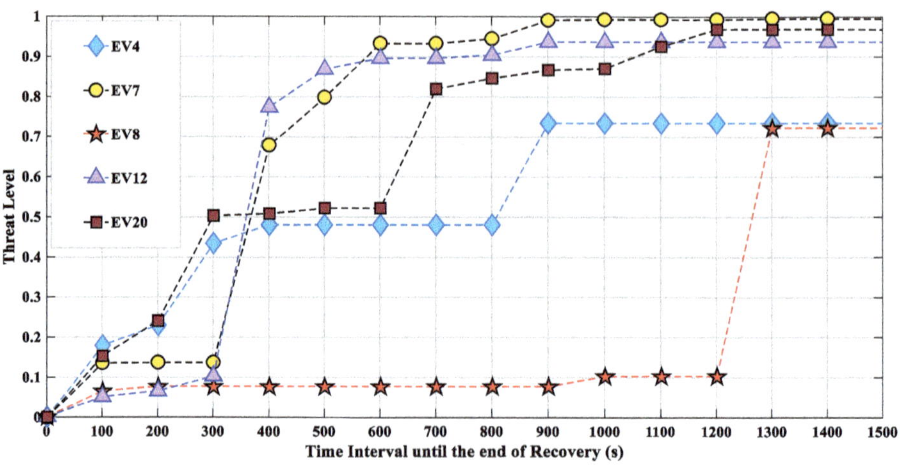

Fig. 5.4 Threat levels with no action

This parameter has a key role in calculations, the threat levels, and thus the optimal decision to temporary isolations. As can be seen, when φ increases, the threat levels enhance more speedily, since the chance of a prosperous attack spread through a

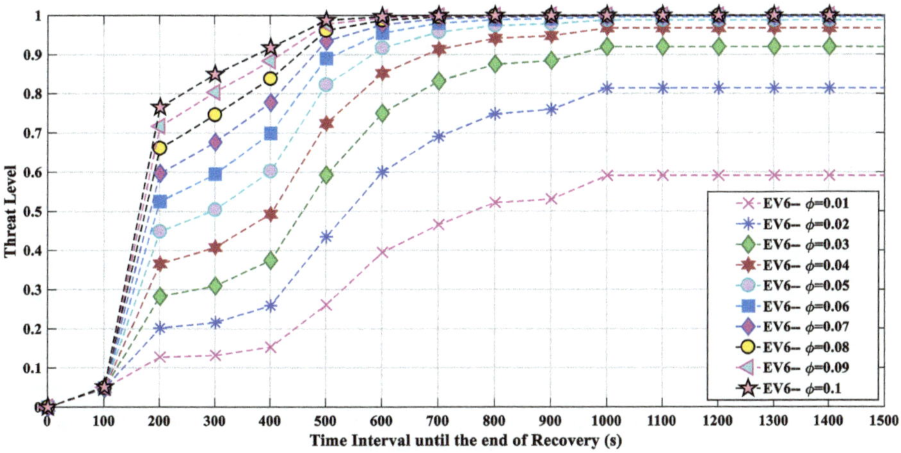

Fig. 5.5 Effect of φ on threat level

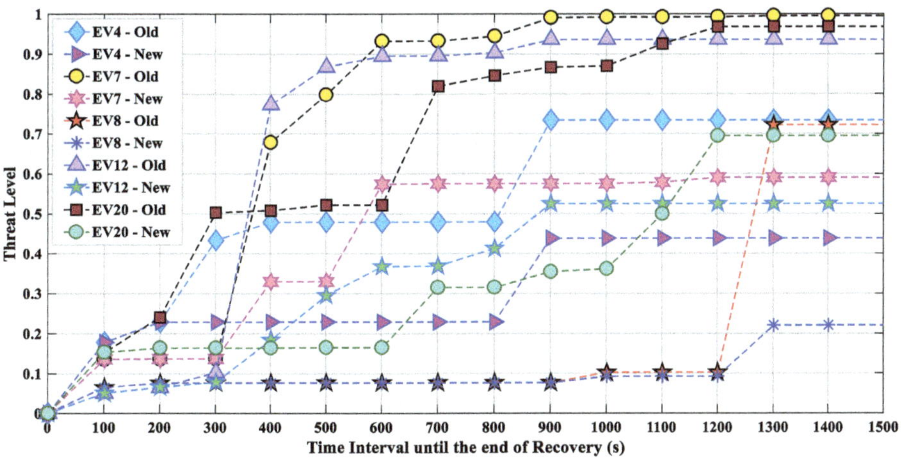

Fig. 5.6 Effect of implementing proposed response model on threat levels

communication relay enhances. In Fig. 5.6, the effect of implementing proposed response model on threat levels is illustrated.

Figure 5.7 illustrates the impact of inspection time on threat levels and attack spread. The assumed data sample space includes $n - \text{EV} = 21$, $\Delta t = 1$ (s), $\varphi = 0.05$, and $\lambda = 1$. We alter $t_{\text{inspection}}$ from 100 to 250 and show several remaining EVs as an example. As mentioned earlier, when a cyberattack occurs on infrastructure, all three detection, inspection, and recovery times have particular importance in countering and responding to the attack. Inspection, which is related to the performance of the system operator for the necessary checks, is an opportunity to spread this cyberattack as much as possible. In the proposed model to mitigate the spread of

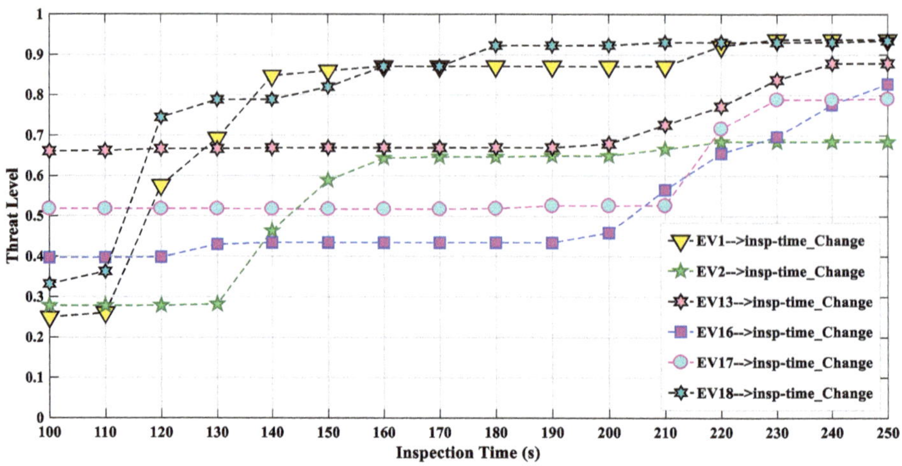

Fig. 5.7 Effect of inspection time on threat levels

attack in the EV network, the values of threat levels at the end of the inspection are used in the optimal decision to temporarily isolate and control the attack until the end of recovery. Therefore, it is obvious that the longer the inspection lasts, the more EVs become infected and the maximum threat level increases also. Hence, the faster inspection completion and the earlier response cause the slower spread of the attack and consequently less damage to the infrastructure.

5.6 Conclusion

The security and resiliency of smart grids are at the risk of cyberattacks due to their well-equipped communication and telecommunication systems and technologies, as well as their interconnected parts and infrastructures. Among the infrastructures that have not been sufficiently studied and are vulnerable to the penetration of cyberattacks is EVI. EVI is an ideal target for such attacks due to the mobile nature of EVs and their shared communication networks. If an EV as the mobile part of the smart grid becomes infected by malware, it can easily spread this attack to other components of the infrastructure like EVSEs or even other equipment of smart grid like PMUs and so on. The mobility pattern of EVs plays a substantial role in the attack spread. In this paper, we proposed a MILP-based response model against cyberattacks in EVs shared communication network that minimizes the maximum threat level and also addresses the EVs and smart grid's interdependency. The results show that this model mitigates the spread of the attack as much as possible and prevents the attack to propagate further into the power grid.

References

1. Su, W., Eichi, H., Zeng, W., & Chow, M. (2012). A survey on the electrification of transportation in a smart grid environment. *IEEE Transactions on Industrial Informatics, 8*(1), 1–10.
2. Mousavian, S., Erol-Kantarci, M., Wu, L., & Ortmeyer, T. (2018). A risk-based optimization model for electric vehicle infrastructure response to cyber attacks. *IEEE Transactions on Smart Grid, 9*(6), 6160–6169.
3. Niyato, D., Kayastha, N., Hossain, E., & Han, Z. (2014). Smart grid sensor data collection, communication, and networking: A tutorial. *Wireless Communications and Mobile Computing (Wiley), 14*(11), 1055–1087.
4. Cheng, X., et al.. (2014). Electrified vehicles and the smart grid: The ITS perspective. *IEEE Transactions on Intelligent Transportation Systems, 15*(4), 1388–1404.
5. Sortomme, E., Hindi, M. M., MacPherson, S. D. J., & Venkata, S. S. (2011). Coordinated charging of plug-in hybrid electric vehicles to minimize distribution system losses. *IEEE Transactions on Smart Grid, 2*(1), 198–205.
6. Mousavian, S., Erol-Kantarci, M., & Ortmeyer, T. (2015). Cyber Attack Protection for a Resilient Electric Vehicle Infrastructure. In *2015 IEEE Globecom Workshops* (*GC Wkshps*), San Diego, CA, pp. 1–6.
7. Abdollahi Biron, Z., Dey, S., & Pisu, P. (2018). Real-time detection and estimation of denial of service attack in connected vehicle systems. *IEEE Transactions on Intelligent Transportation Systems, 19*(12), 3893–3902.
8. Siegel, J. E., Erb, D. C., & Sarma, S. E. (2018). A survey of the connected vehicle landscape—Architectures, enabling technologies, applications, and development areas. *IEEE Transactions on Intelligent Transportation Systems, 19*(8). [8]. Koscher, K., et al. (2010). Experimental security analysis of a modern automobile. In *2010 IEEE symposium on security and privacy*, Berkeley/Oakland, CA, pp. 447–462.
9. Morrison, G. S. (2018). Threats and mitigation of DDoS cyberattacks against the U.S. power grid via EV charging. Electronic Thesis or Dissertation, Wright State University.
10. Fraunhofer. (2018). Expert from Fraunhofer ITWM uncovers security vulnerabilities of charging stations. [Online]. Available: https://www.fraunhofer.de/en/press/research-news/2018/January/security-vulnerabilities-of-charching-stations.html
11. Alcaraz, C., Lopez, J., & Wolthusen, S. (2017). OCPP protocol: Security threats and challenges. *IEEE Transactions on Smart Grid, 8*(5), 2452–2459.
12. Engebretson, P. (2013). *The basics of hacking and penetration testing: Ethical hacking and penetration testing made easy* (2nd ed.). Syngress.
13. Mrabet, Z. E., Kaabouch, N., Ghazi, H. E., & Ghazi, H. E. (2018). Cyber-security in smart grid: Survey and challenges. *Computers and Electrical Engineering, 67*, 469–482.
14. Knapp, E. D., & Samani, R. (2013). *Applied cyber security and the smart grid: Implementing security controls into the modern power infrastructure* (1st ed.). Syngress.
15. Shapsough, S., Qatan, F., Aburukba, R., Aloul, F., & Al Ali, A. R. (2015). Smart grid cyber security: Challenges and solutions. In *2015 International Conference on Smart Grid and Clean Energy Technologies (ICSGCE)*, Offenburg, pp. 170–175.
16. NHTSA. (2019). Vehicle-to-Vehicle Communication. [Online]. Available: https://www.nhtsa.gov/technology-innovation/vehicle-vehicle-communication
17. Chan, A. C., & Zhou, J. (2013). On smart grid cybersecurity standardization: Issues of designing with NISTIR 7628. *IEEE Communications Magazine, 51*(1), 58–65.
18. Li, Y., Wang, R., Wang, P., Niyato, D., Saad, W., & Han, Z. (2012). Resilient PHEV charging policies under price information attacks, In *2012 IEEE Third International Conference on Smart Grid Communications* (*SmartGridComm*), pp. 389–394.
19. Carryl, C., Ilyas, M., Mahgoub, I., & Rathod, M. (2013). The PEV security challenges to the smart grid: Analysis of threats and mitigation strategies. In *2013 International Conference on Connected Vehicles and Expo (ICCVE)*, pp. 300–305.

20. Chan, A. C., & Zhou, J. (2014). Cyber-physical device authentication for the smart grid electric vehicle ecosystem. *IEEE Journal on Selected Areas in Communications, 32*(7), 1509–1517.
21. Su, H., Qiu, M., & Wang, H. (2012). Secure wireless communication system for smart grid with rechargeable electric vehicles. *IEEE Communications Magazine, 50*(8), 62–68.
22. Mousavian, S., Valenzuela, J., & Wang, J. (2015). A probabilistic risk mitigation model for cyber-attacks to PMU networks. *IEEE Transactions on Power Systems, 30*(1), 156–165.
23. Zonouz, S., Davis, C. M., Davis, K. R., Berthier, R., Bobba, R. B., & Sanders, W. H. (2014). SOCCA: A security-oriented cyber-physical contingency analysis in power infrastructures. *IEEE Transactions on Smart Grid, 5*(1), 3–13.
24. Esmalifalak, M., Han, Z., & Song, L. (2012). Effect of stealthy bad data injection on network congestion in market based power system. In *2012 IEEE wireless communications and networking conference* (*WCNC*), Paris, France, pp. 2468–2472.
25. U.S. Department of Energy Office, Electricity Subsector Cybersecurity Risk Management Process. [Online]. Available: Cybersecurity Risk Management Process (RMP) Guideline- Final (2012, May).
26. Esmalifalak, M., Liu, L., Nguyen, N., Zheng, R., & Han, Z. (2017). Detecting stealthy false data injection using machine learning in smart grid. *IEEE Systems Journal, 11*(3), 1644–1652.
27. Altunay, M., Leyffer, S., Linderoth, J. T., & Xie, Z. (2011). Optimal response to attacks on the open science grid. *Computer Networks, 55*(1), 61–73.
28. Chen, N., Wang, M., Zhang, N., & Shen, X. (2020). Energy and information management of electric vehicular network: A survey. *IEEE Communications Surveys & Tutorials, 22*(2), 967–997, Second quarter.

Chapter 6
Electric Vehicle Services to Support the Power Grid

Saleh Aghajan-Eshkevari, Mohammad Taghi Ameli, and Sasan Azad

6.1 Introduction

The transportation sector is known as one of the world's largest energy consumption sectors, which is about one-third of the total energy consumed in this sector alone. On the other hand, vehicles based on internal combustion engines have taken the leading share in the current transportation system. Hence, the existing transportation system has a significant role in air pollution and greenhouse gas emissions. Therefore, to reduce the environmental issues caused by fossil fuel vehicles and less dependence on fossil fuels, electrification of the transportation system is considered a suitable solution. On this basis, EVs, regarded as a clean and environmentally friendly alternative to fossil fuel vehicles, are under special attention [1, 2].

Due to the high production cost of EVs and insufficient charging equipment, the growth of EVs penetration in the transportation system is slow. Policies and regulations have been adopted worldwide to extensively overcome obstacles and develop EVs. According to the International Energy Agency, the whole number of EVs worldwide will catch 130 million by 2030 [3].

Although this large number of EVs integrated into the transportation system can diminish greenhouse gas emissions, it is predictable that the uncoordinated integration of these EVs in the power grid can cause many problems such as power quality reduction and overloading transmission lines and transformers, voltage drop, etc. Also, it can affect the stability and security of the power network [3, 4]. Uncoordinated charging is when the EVs start charging immediately after plugging into the grid, regardless of the grid status. And the charging process stops when the

S. Aghajan-Eshkevari · M. T. Ameli (✉) · S. Azad
Department of Electrical Engineering, Shahid Beheshti University, Tehran, Iran
e-mail: s.aghajaneshkevari@mail.sbu.ac.ir; m_ameli@sbu.ac.ir; sa_azad@sbu.ac.ir

V. Vahidinasab, B. Mohammadi-Ivatloo (eds.), *Electric Vehicle Integration via Smart Charging*, Green Energy and Technology,
https://doi.org/10.1007/978-3-031-05909-4_6

EV battery is fully charged, or the EV is plugged out from the grid. Therefore, in the case of uncoordinated charging, EVs are considered a new type of electrical load for the power grid. However, with the increase in the number of EVs in the transportation system, this electrical load type also increases. Also, due to the spatial and temporal uncertainty of EVs' charging load in the power grid, accurate forecasting and control of these loads will be challenging. Furthermore, with the growth of EVs in the distribution network, power flow will change significantly. The simultaneous charge demand of many EVs in a particular location of the network causes congestion in transmission lines, resulting in voltage drops and line loss. Also, in addition to the problems that arise with the uncoordinated charging of EVs in controlling and operating the power grid, this charging method causes overload in transformers and accelerates their aging. Therefore, uncoordinated charging also has significant adverse effects on power grid equipment [3–6].

Therefore, in order to reduce the harmful effects of uncoordinated charging, various strategies have been presented for the charging and discharging coordination of EVs. In addition to facilitating the adverse impacts of uncoordinated integration in the grid, these strategies can increase network capacity to host more EVs. Also, by optimal charging and discharging management of EVs in the power network and considering the uncertainties associated, such as plug-in and plug-out time from the grid, initial and final state of charge (SOC) of the EV battery, and EV's energy requirement for next trip, EVs can be used to improve grid operation [5, 6]. In fact, using smart grid capabilities, EVs are not only considered as a type of electrical load but also as a distributed energy storage system that can provide services to the grid. Hence, vehicle-to-grid (V2G) technology can facilitate bidirectional energy exchange between EVs and the power network [5–8]. In addition, batteries are more efficient in comparison with other energy storage technologies. Although EVs are less available to the network operator than fixed batteries, initial investment cost in fixed batteries is high. In fact, the primary purpose of purchasing EVs is transportation, and providing services to the grid is a secondary priority. However, since EVs are parked for about 95% of the day [9], they can also provide services so that EV owners can earn financial benefits. However, due to the insufficient capacity of a single EV, its charge and discharge control is not performed directly by the network operator. Usually, the EVs are divided into several clusters, and an aggregator performs each EV group's charging and discharging control. Therefore, as shown in Fig. 6.1, the aggregator has the responsibility of managing the charge and discharge of EVs to provide services to the power grid.

So far, many investigations have been done for optimal management of charging and discharging of EVs to provide services to the network, and this issue is still of particular interest to researchers. EVs can provide various services by injecting or absorbing active and reactive power, such as frequency regulation [10], voltage regulation [11], peak shaving [12], valley filling [13], loss reduction [14], etc. Therefore, they can improve network performance properly. Also, by smart charging of EVs during renewable energy production peak hours, it is possible to support the renewable energy sources integration in the power network and help the penetration

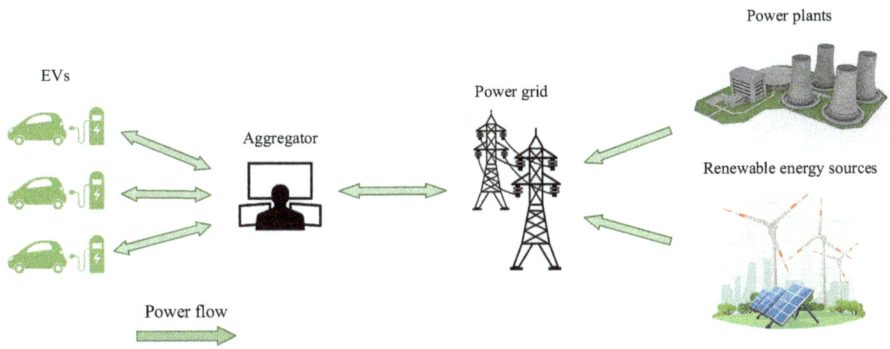

Fig. 6.1 EVs' power exchange with smart grid

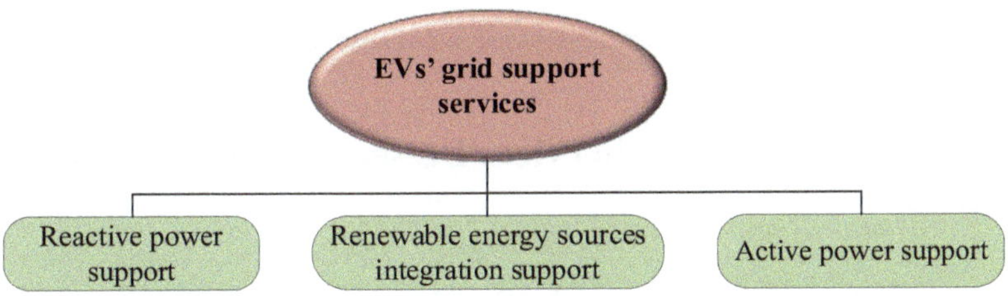

Fig. 6.2 EVs' services to support power system operation

level increment of these energy sources in the grid [15]. Thus, according to Fig. 6.2, in this chapter, the services presented to the electricity network by EVs optimal charging and discharging are classified into three categories, namely active power support services, reactive power support services, and services to support renewable energy sources integration [5, 6]. Services like frequency regulation and peak shaving are classified in the active power support services category. Services that can be provided by injecting or absorbing reactive power, such as voltage control and loss reduction, are listed in the reactive power support services category. And support services for integrating solar and wind energy sources in the power grid are incorporated in the renewable energy integration support category.

Therefore, in this chapter, the various services of EV are first classified into three categories: active power support, reactive power support, and support for renewable energy integration, and the services related to each category are analyzed in detail. Then, to better support the power grid, the ability of EVs to provide a combination of different services to the grid is discussed. After that, mathematical modeling of EVs' charging and discharging optimization problem in the power grid is presented. Finally, the current status, challenges, and outlook of smart charging of EVs are considered. In general, the topics discussed in this chapter are summarized in Fig. 6.3.

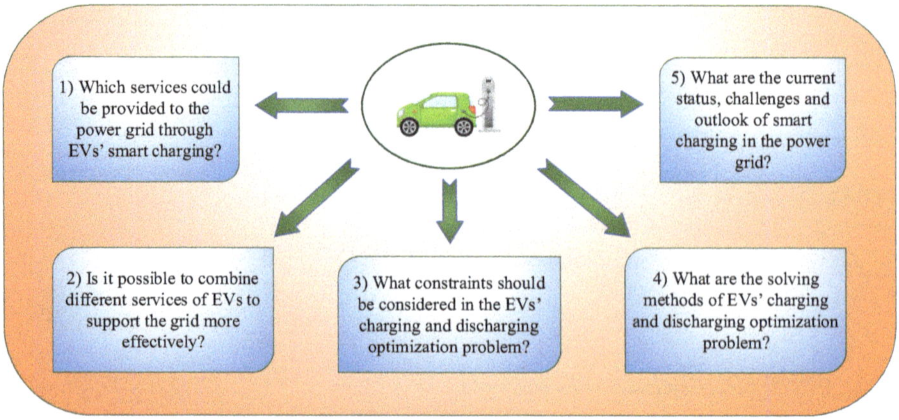

Fig. 6.3 Graphical abstract of the topics discussed in this chapter

6.2 Classification of EV Services Presentable to the Power Grid

Although increasing the penetration level of EVs in the electricity network poses many new challenges for the network operator in control and operation of the power system, EVs can create new opportunities by providing services to the network. This section considers various services provided to the power grid with optimal charging and discharging management of EVs. In general, EV services can be classified into three categories: active power support, reactive power support, and support for renewable energy integration. In the following, each of these categories is considered in detail.

6.2.1 EV's Active and Reactive Power Support Services

A review of various studies in the field of EV charging and discharging management shows that EVs have been used to provide multiple services related to active and reactive power support. As shown in Fig. 6.4, EV's active power support services include frequency regulation, load variance minimization (peak shaving and valley filling), loads restoration, loss reduction, and voltage control.

However, the main target of reactive power support by EVs is to provide voltage control service in the power network. Several articles have shown that the grid voltage can be maintained at an acceptable range by controlling reactive power. On the other hand, EV chargers have the capability to inject inductive or capacitive reactive power into the power network by appropriately adjusting the current phase angle. By reactive power compensation, EVs not only help to control the grid voltage [11] but also can reduce grid losses [14] and improve the power factor

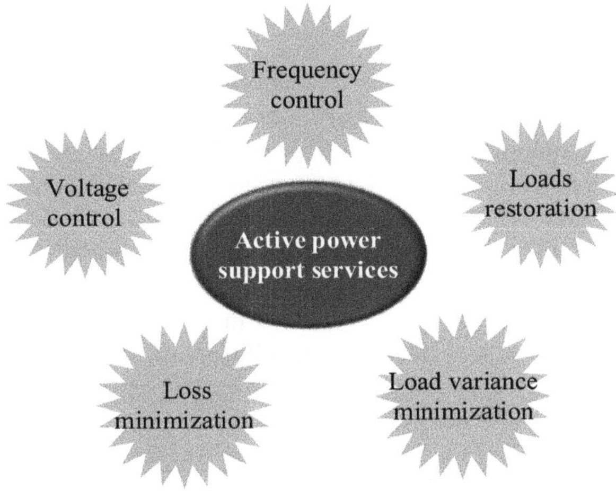

Fig. 6.4 EVs' active power support services

of the electricity network [16]. Also, because the reactive power compensation service is provided by the DC-link capacitor of the bidirectional EV battery charger, this service does not damage the EV battery because the battery is not directly responsible for providing this service. Therefore, with the development of chargers that can control the reactive power of EV, this service can be beneficial.

In the following, we will consider the capabilities of EVs in providing each of these services to the power grid. However, it should be mentioned that EVs are not always available, unlike typical energy storage systems, which are always available to the system operator. There are uncertainties associated with EVs that should be regarded in charge and discharge optimization problems of EVs. Uncertainties related to the EV include arrival time, departure time, and the distance traveled by the vehicle during the day, which are usually modeled using appropriate probability distribution functions. Therefore, to model the behavior of EV more accurately, these uncertainties must be considered in charging and discharging optimization problems.

6.2.1.1 Frequency Control

An imbalance between the generation and demand of electrical power causes frequency fluctuations in the power system. More generation than demand increases the frequency, and less generation than demand decreases the frequency of the power network. Both mentioned cases create many problems for the system operator and can even jeopardize the power system's stability. Frequency control is categorized into three classes named primary, secondary, and tertiary frequency control based on different factors like time of responding, droop control, and power necessities. Traditionally, power plants provide frequency control service to the power system. However, EV batteries have recently been employed to regulate the power grid frequency because of their high response rate. Also, EVs can act as both a generator and a power consumer in the power grid, and therefore have the ability to participate in both regulation-down and regulation-up services. However, it should be considered that discharging the EV battery in the grid will cause battery degradation. Therefore, EV owners should be encouraged to participate in frequency regulation service with appropriate financial incentives [17].

Due to the high capability of EVs to provide frequency regulation service, several studies have investigated the economic and technical issues related to the participation of EVs in frequency control [10, 18–22]. In [10], unlike most researches that use EVs parked in charging stations to provide frequency control service, in this paper, the capability of individual EVs to provide frequency control service is analyzed, and the charging demand of each EV is also considered. In [20], an optimal strategy is presented for the EVs parked in the charging station to take part in the secondary frequency control service, which also regards the satisfaction of EV charging requirements. Also, for fair dispatching allocation from the control center between EVs according to their charging need, two optimal strategies for real-time based on area control error (ACE) and area regulation requirement (ARR) are proposed. These two strategies are implemented on a two-area interconnected power system. The simulation results show that these two strategies can reduce the frequency deviation of the system by 12.66% and 16.78%, respectively. In [21], EV participation in frequency control is performed with the consideration of battery degradation and with the aim of reducing the EV owner's charging cost. Also, to increase the profit of EV aggregators, the remaining capacity of the EV charger is utilized for reactive power injection into the grid and providing voltage control service. However, in [22], EVs charge and discharge coordination is done to participate in the primary frequency control service. In this reference, a two-level frequency control method based on consensus is proposed. In the higher level of the proposed method, the power system frequency deviation decreases, and in the lower level, the cost of frequency control and the degradation of the EV battery is minimized. Also, graph theory is applied to establish a communication network between control areas in the vicinity at the higher level and between adjacent EVs at the lower level.

6.2.1.2 Load Variance Minimization, Peak Shaving, and Valley Filling

Reducing the peak load of the power grid by controlling the loads is called peak shaving, but creating loads in the grid during non-peak periods is called valley filling. However, load variance minimization is performed to flatten the grid load curve, in which the grid load decreases during the peak period and increases during the non-peak period. In fact, load variance minimization service is a combination of peak shaving and valley filling services and has the capabilities of both services. By providing appropriate incentives, the EVs charging demand can be transmitted from peak to off-peak times, which leads to congestion prevention in network transmission lines and also loss reduction. Equations (6.1, 6.2, and 6.3) represent the objective functions related to peak shaving [12], valley filling [5], and load variance minimization [23], respectively, which are utilized in EVs charging and discharging optimization problems.

$$\min \sum_{t=1}^{T} \left(P_{\text{Peak}.t} - P_{\text{target}.t}\right) \tag{6.1}$$

$$\min \sum_{t \in T_{\text{off-peak}}} \left(P_{\text{target}.t} - P_{\text{base}}\right) \tag{6.2}$$

$$\min \sum_{t=1}^{T} \left(P_{c.t} * N_{c.t} + P_{\text{Load}.t} - P_{d.t} * N_{d.t} - \overline{P}\right)^2 \tag{6.3}$$

In Eq. (6.1), $P_{\text{Peak},t}$ and $P_{\text{target},t}$ indicate the load peak power and target peak power at time step t, respectively. In (6.2), P_{base} indicates the base load of the grid. In (6.3), $P_{\text{c,t}}$ and $P_{\text{d,t}}$ indicate the EVs charge and discharge power at time *t*, respectively, and $P_{\text{Load,t}}$ and $\overline{P}$ are basic load at time *t* and average load during a day without EVs, respectively.

A review of various researches shows that EVs charging and discharging management aiming at minimizing the load variance [23–27], peak shaving [12, 28–29], valley filling [13, 28] has been considered by researchers [12] presents an energy management strategy based on multi-objective optimization to achieve peak shaving, loss minimization, and maximizing EV owner's profit. In [13], a centralized strategy for EVs charging on a large scale is presented to fill the valleys. Two indexes are defined as the index of capacity margin and the index of charging priority. The first index is employed to choose the time interval when the power network has much surplus power for charging EVs. The second index is employed to decide EVs' charging priority in each interval. In [24], optimal scheduling of EVs charging and discharging is performed for load variance minimization. In the presented method, the EVs start charging when the power system loading is less than the target loading, and the EVs are discharged when the power network loading is more than the target loading. Thus, the efficiency of the presented method is considerably dependent on the capability of the grid-connected EVs and the adjustment of the target load and the grid load. In [25], a method for smart charging of EVs in the distribution system is presented. The proposed smart charging method has been implemented locally in each EV, which helps to improve system scalability and reliability. The presented algorithm of this paper considers the goals of minimizing both the load variance and the EVs charging cost. Voltage imbalance is also diminished by charging process coordination between the system phases.

6.2.1.3 Loads Restoration

Due to the vital role of electrical energy in human life, power outages due to natural disasters or other factors can cause significant damage and even threaten the lives of hospital patients. Accordingly, the efficient and rapid restoration of power system loads after power outages is essential to the system operator. On the other hand, EVs can be thought of as distributed energy storage systems. Therefore, considering the V2G capability, the energy stored in their batteries can be utilized to recover critical loads of the power network. However, the ability of EVs to restore the grid loads has taken little notice in research [30–32]. In [30], a restoration strategy based on mixed-integer linear programming (MILP) is presented that aims to maximize the restored loads in the shortest time by optimizing the load restoration sequences and charging and discharging of EVs. The results of this investigation show that EVs can help to restore more loads. In [31], EVs parked in public and residential parking lots have been employed to recover loads during blackouts. Also, the technical and economic benefits of EVs participating in microgrid formation for critical loads restoration have been evaluated. In [32], EVs are utilized to balance generation and demand in order to achieve frequency regulation during the restoration service. In fact, the charge and discharge of EVs are optimized to control the system frequency during restoration. Then, at the end of the restoration service, the SOC of the batteries returns to the initial state.

6.2.1.4 Loss Minimization

Loss Minimization by Active Power Management

Loss in the power system is caused by the loss of energy in the form of heat in conductors. Usually, energy losses increase as demand increases. Therefore, uncoordinated charging of EVs, in addition to increasing the peak demand of the network, will lead to rising losses. Therefore, in the EVs charging and discharging coordination problem, minimizing network losses has also been considered. Loss minimization can be attained by controlling EVs' active or reactive power. In this section, loss minimization strategies with active power control are reviewed. Eq. (6.4) expresses the objective function for minimizing grid loss [12]. In (6.4), N_{line} indicates the number of gridlines and $I_{l,t}$ and R_l indicate the current and resistance of line l respectively.

$$\min \sum_{t=1}^{T} \sum_{l=1}^{N_{\text{line}}} R_l * I_{l.t}^2 \tag{6.4}$$

Active power control of EVs to decrease system power loss is studied in [12, 26, 33–36]. In [33], a coordinated EV charging strategy is presented considering the spatial and temporal uncertainties that minimize distribution network loss and voltage deviations. In [34], multi-objective optimization for EVs and distributed generation coordination in the distribution grid is proposed, which evaluates several power quality-related indexes like neutral current, voltage unbalance, loss of energy, and bus voltage in the optimization problem. The presented method is tested in an Australian low voltage distribution network. The simulation results show that the energy loss in the coordinated operation mode is reduced by 59.45% compared to the uncoordinated mode. In [35], unlike other references, network loss is diminished by optimal managing of EVs' both active and reactive power.

Loss Minimization by Reactive Power Control

As mentioned, by controlling the reactive power of EVs, network loss can be minimized, which in turn alleviates the network operation cost. In [14, 35, 37–39], loss minimization by EV reactive power compensation has been investigated. In [14], a two-layer smart energy management strategy is presented for active and reactive power control of EVs in the distribution network. In the first layer, active power management of EVs is performed from the nodal aggregator's perspective, which aims to minimize the total daily cost. In the second layer, the reactive power management of the EVs is performed from the distribution network operator's point of view to minimize grid power loss. In [38], reactive power management of EVs is adopted to diminish the reactive power support of conventional generators, leading to the generators' operation at higher power factor and increasing financial profits, decreasing loss, and improving voltage stability of the power system.

6.2.1.5 Voltage Control

Voltage Control by Active Power Management

While connecting generating or consuming units of electrical energy to the distribution network, voltage issues will occur, leading to an increase or decrease in voltage. Uncoordinated charging of EVs in the network may also increase loading and voltage drop problems during peak periods. Therefore, different methods have been developed to overcome voltage issues by utilizing various devices and sources such as distributed generation units, on-load tap changers, and capacitor banks. Furthermore, voltage issues can be mitigated by optimal management of the active or reactive power of EVs. In this section, voltage regulation by active power management of EVs is considered. Voltage control can be implemented by defining constraints for the voltage range, as shown in Eq. (6.5), where V_i, $V_i^{\min}$, and $V_i^{\max}$ indicate the voltage domain, the minimum and the maximum voltages at node i, respectively. Also, voltage control can be carried out by defining the objective function to minimize the voltage deviation of the grid buses [5] or to minimize the grid voltage imbalance [34], which are the two objective functions in Eqs. (6.6) and (6.7), respectively.

$$V_i^{\min} \leq V_i \leq V_i^{\max} \tag{6.5}$$

$$\min \sum_{i=1}^{N_{\text{bus}}} (V_r - V_i)^2 \tag{6.6}$$

$$\min \sum_{t=1}^{T} \sum_{i=1}^{N_{\text{bus}}} \left| \frac{V_{t.i}^{-}}{V_{t.i}^{+}} \right| \tag{6.7}$$

In (6.6), V_r and N_{bus} indicate the voltage reference and the number of grid nodes, respectively. In (6.7), $V_{t.i}^{-}$ and $V_{t.i}^{+}$ indicate the negative and positive sequence voltage at time t and bus i, respectively.

Achieving voltage control by active power management of EVs has been investigated in [27, 33–35, 40–43]. In [41], the impact of EVs on the voltage imbalance of power grid is regarded. In this paper, by optimal selection of three factors, namely charging or discharging EV, EV's connection point among grid phases, and the rated power of EV's charging or discharging, the voltage unbalance factor (VUF) is minimized. The effect of coordinated and uncoordinated charging of EVs on the VUF is also evaluated. In [43], EV's energy management and active power curtailment of photovoltaic (PV) arrays are proposed with the aim of voltage control in low voltage distribution networks with high penetration level of PV sources. Coordinated operation of EVs and PV also reduces the required curtailment of active power from PV arrays. Furthermore, a distributed control strategy consisting of two consensus algorithms is employed to use the restricted capacity of EVs battery effectively.

Voltage Control by Reactive Power Management

Utilizing the EV's capability for injection or absorption of reactive power to control grid voltage has been studied in [11, 21, 29, 35, 38, 39, 42]. In [11], using electric vehicle chargers (EVCs), a two-level strategy for voltage control in distribution grids is presented. At the upper level, the voltage control is carried out by measurement of the critical buses voltages, and the resulted control signals are distributed to all EVCs. Each EVC's active and reactive power output is decided at the lower level regarding the EV's operating limitations and charging needs. In [39], both active and reactive power management is done to reduce the energy loss cost and transformers operation. Also, the proposed method in this article can enhance power quality-related parameters such as voltage and power factor by appropriate management of reactive power. In [42], active and reactive power control of EVs is executed for minimizing the cost of purchased energy by the distribution network operator (DNO) and minimizing the grid voltage deviations. Also, linearization techniques are employed to convert the nonlinear objective function and constraints to the MILP model.

6.2.2 *Support Services for Renewable Energy Sources Integration*

In order to address environmental issues, the penetration level of renewable energy sources (RESs) is rapidly increasing in the power network. However, since the output power of RES fluctuates widely, the high penetration of these resources into the power system creates problems in system operation. A suitable solution to overcome these problems and reduce the curtailment of renewable energy sources is the joint operation of EVs and RESs [5, 6]. EVs using V2G capability can effectively support the integration of RESs into the power grid. Therefore, EVs charge their batteries in periods of high RESs generation. Also, in periods of low RESs generation, they meet the demand and support to increase the RESs penetration in the grid by discharging their batteries into the network.

EVs' charge and discharge coordination aim to increase the penetration of PV systems [15, 40, 43–45] wind energy sources [15, 36, 46, 47] in the power grid has been studied in several investigations. In [44], a charging management scheme for EV is presented to coordinate a quantity of self-consumption of PV output through shifting the EVs charging period, diminishing PV curtailment induced by voltage rise in the low voltage distribution grid. Also, the auction mechanism is introduced to ensure both the equity of benefit in each customer and the freedom that allows customers to participate in the EV charging management scheme voluntarily. In [15], a smart charging strategy for EVs is proposed to decrease the effect of solar and wind power fluctuations by regarding their temporal and spatial features. This study modulates EVs as demand-responsive loads by introducing stochastic dynamic pricing. Then, two indices are determined to measure the fluctuations of renewable energies output, and the cost of EV charging is also considered as an economic indicator. These indicators are minimized in an optimal charging model. In [46], a coordinated scheduling model is proposed to improve the absorption of wind power in the power system. The proposed model is formulated as a multi-objective optimization problem and includes thermal units, wind units, and EV charging stations. In this paper, by optimizing the output power of thermal units, the absorbed wind power, and the power of charging stations, the goals of reducing wind power curtailment and both generation and emission costs are attended. The results show that EVs can improve wind power absorption.

Table 6.1 shows the presented categorization of EV services to support power system operation.

6.3 Combination Capability of EVs' Different Services

To better and more effectively support the performance of the power system, in many studies, different EV services have been combined. Thus, EVs offer several services according to the needs of the system. In [21], an active power support

Table 6.1 Classification of EV services provided to the power system

	Services	Reference
	Frequency control	[10, 18–22]
	Load variance minimization, peak shaving, and valley filling	[12, 13, 23–29]
Active power support	Loads restoration	[30–32]
	Loss minimization by active power management	[12, 26, 33–36]
	Voltage control by active power management	[27, 33–35, 40–43]
Reactive power support	Loss minimization by reactive power management	[14, 35, 37–39]
	Voltage control by reactive power management	[11, 21, 29, 35, 38, 39, 42]
Renewable energy sources integration support	Solar and wind sources integration support	[15, 36, 40, 43–47]

service, i.e., frequency control, is combined with a reactive power support service, i.e., voltage control. In this paper, the priority is with frequency regulation. Then the remaining capacity of EV chargers is utilized to exchange reactive power with the grid and provide voltage control service. Active and reactive power support services can be combined effectively, but prioritization for services must be considered. The combination of peak shaving services and frequency tuning can also be done to plan the day ahead. The peak shaving service is delivered to the network at a specific time period, and EVs can be used to provide frequency tuning service during the remaining time. However, these services may be needed simultaneously for real-time applications, which makes it challenging to combine these services.

Similarly, combining frequency control service and renewable energy support is possible for day-to-day planning because the renewable energy support service is time-dependent. At certain times EVs can be used to provide this service, and at other times they can be used to provide frequency control service. However, combining these two services in real-time is associated with difficulties [5].

6.4 Mathematical Modeling of EVs' Charging and Discharging Optimization Problem in the Power System

The problem of optimizing the charge and discharge of EVs includes various objective functions and constraints, and on a large scale, it has a large number of decision variables. Hence, this issue is usually complex. This section presents the mathematical formulation of some of the main constraints of the charging and discharging optimization problem. And then, different models of charge and discharge optimization problems, including linear programming (LP), nonlinear programming (NLP), mixed-integer quadratic programming (MIQP), MILP, and Mixed-Integer Nonlinear Programming (MINLP), are introduced. Finally, the methods and algorithms used in various researches to solve this optimization problem are reviewed.

6.4.1 Constraints on EVs' Charging and Discharging Optimization Problem

In Sect. 6.2, some objective functions are formulated to provide EV services to the grid. However, restrictions must also be considered in charge management and discharge of EVs in the power grid [3]. In general, the constraints of this issue can be divided into two categories: (1) EV-related constraints and (2) network-related constraints. In the following, some of the main constraints are examined. It should be noted that depending on the type of problem, other constraints may be used in formulating the problem.

6.4.1.1 EV Constraints

By reviewing different articles, it can be observed that in the case of EVs, there are usually four constraints in formulating the problem of charge and discharge optimization:

1. Constraint of charging and discharging power of the EV: Charging and discharging the EV battery with excessive power accelerates the battery's aging. Each type of EV, depending on its battery, has a virtual range for charging and discharging. In Eqs. (6.8) and (6.9), EVs' allowable range of charging and discharging power is applied, respectively. In these equations, $P_{t,n}^{\text{ch}}$ and $P_{t,n}^{\text{dis}}$ are the charge and discharge power of the nth EV at time t, respectively. $P_{\min,n}^{\text{ch}}$ and $P_{\max,n}^{\text{ch}}$ represent the minimum and maximum charge power, respectively, and $P_{\min,n}^{\text{dis}}$ and $P_{\max,n}^{\text{dis}}$ also represent the minimum and maximum EV discharge power. $x_{t,n}^{\text{ch}}$ and $x_{t,n}^{\text{dis}}$ are also binary variables.

$$P_{\min,n}^{\text{ch}} * x_{t,n}^{\text{ch}} \leq P_{t,n}^{\text{ch}} \leq P_{\max,n}^{\text{ch}} * x_{t,n}^{\text{ch}} \tag{6.8}$$

$$P_{\min,n}^{\text{dis}} * x_{t,n}^{\text{dis}} \leq P_{t,n}^{\text{dis}} \leq P_{\max,n}^{\text{dis}} * x_{t,n}^{\text{dis}} \tag{6.9}$$

2. Lack of synchronization of charging and discharging constraint: To prevent the occurrence of synchronous charge and discharge mode, a constraint is defined as follows:

$$x_{t,n}^{\mathrm{ch}} + x_{t,n}^{\mathrm{dis}} \le 1 \tag{6.10}$$

3. EV battery capacity constraint: The amount of charge and discharge depth of the EV battery affects its life. Excessive discharge depth increases battery damage, and overcharging the battery may cause the battery to wear out. Therefore, the battery SOC according to Eq. (6.11) should be within a certain range [3]. In (6.11), $\mathrm{SOC}_{t,\,n}$ represents the SOC of nth EV in time t. $\mathrm{SOC}_{\min,\,n}$ and $\mathrm{SOC}_{\max,\,n}$ indicate the minimum and maximum allowable SOC of nth EV, respectively.

$$\mathrm{SOC}_{\min,n} \le \mathrm{SOC}_{t,n} \le \mathrm{SOC}_{\max,n} \tag{6.11}$$

4. EV owner's charging requirement constraint: This constraint is one of the essential constraints in the charging and discharging management problem of EVs that must be considered because the primary purpose is to meet the EV owner's charging requirement for transportation. Other goals, such as providing service to the network, are the next priority. Therefore, the SOC of the EV battery should reach the desired value of the EV owner when plugging out from the network. In Eq. (6.12), this constraint is stated [3].

$$\blacksquare_n \Delta t \sum_{t=\Gamma_n^s}^{\Gamma_n^e} P_{t.n}^{\mathrm{EV}} = \left(\mathrm{SOC}_n^e - \mathrm{SOC}_n^s \right) C_n \tag{6.12}$$

In (6.12), η_n represents the charger efficiency of the nth EV. Γ_n^s and Γ_n^eindicate the plug-in and plug-out time of nth EV from the network, respectively. SOC_n^s and SOC_n^e indicate the SOC of nth EV at plug-in and plug-out time from the grid, respectively. C_n represents the battery capacity of the nth EV.

6.4.1.2 Network Constraints

In various articles, various network-related constraints have been considered. In this section, two of the most frequent constraints in the articles are stated:

1. Bus voltage constraint: Determining the allowable range for the voltage of the bus of the power system is necessary to maintain the system's stability. Therefore, in EV charge and discharge management problems, a constraint according to Eq. (6.5) is considered for the voltage of the nodes.
2. Transmission power of lines constraint: To prevent congestion and overload of transmission lines, the transmission power of lines must be less than their thermal

limit. This constraint is shown in Eq. (6.13) [5]. In this regard, P_l and $P_{L\max}$ are the transfer power and heat capacity of line l, respectively.

$$-P_{L\,\max} \leq P_l \leq P_{L\,\max} \tag{6.13}$$

6.4.2 *Mathematical Models and Problem-Solving Methods for Optimizing Charge and Discharge of EVs*

As mentioned earlier, the mathematical modeling of EVs' charging and discharging optimization problem usually include the LP, NLP, MILP, MIQP, and MINLP models. In general, the LP model provides a simple and efficient model for optimizing the charge and discharge of EVs, which can be solved using mathematical optimization methods such as the simplex method. With the help of the linear model, the optimal global solution to the charge and discharge optimization problem can be achieved. Still, this model is limited to the objective function and linear constraints. In addition, the simplex method is not efficient enough to solve more complex nonlinear models, and hence, other methods and algorithms have been developed to solve these models.

Nonlinear models can be divided into two categories: convex optimization models and non-convex optimization models. In general, convex models have less solution time than non-convex models because convex models can be solved using the internal point algorithm. Still, there is almost no complete and robust algorithm for solving non-convex models [3]. Hence, in many investigations, the meta-innovative algorithms such as genetic algorithm (GA) [14, 15], particle swarm optimization (PSO) [38], differential evolution (DE) [14, 34], the improved quasi-electromagnetic algorithm (IEMA) [12] have been used to solve complex models. The main advantage of meta-innovative algorithms is that they have less computational time than traditional optimization methods but usually can't achieve the optimal global response.

6.5 Current Status, Challenges, and Outlook

As previously discussed, smart charging and discharging of EVs can provide a variety of services to the power grid and help the improvement of network performance. Therefore, in several projects worldwide, the capability of EVs to provide different services to the grid has been tested. For example, in 2014, the Southwest Research Institute deployed an aggregation of electric delivery trucks to manage charge activities to provide frequency regulation service within the Electric Reliability Council of Texas (ERCOT). In this project, the EVs' charging is suspended when the power grid frequency drops below a certain threshold. Hence, the vehicles did not inject power into the grid, and just their charging process was postponed [48]. Another electric vehicle smart charging implementation has been conducted by Mitsubishi in the Netherlands. In this project, a fleet of Mitsubishi Outlander PHEVs located in residential and workplace parking sites participate in frequency regulation. Also, this fleet provides spinning reserve service to the power grid. While implementing this project in 2017, there were more than 25,000 Mitsubishi Outlander PHEVs on the road in the Netherlands [49, 50].

Although smart charging and discharging of EVs can improve power grid operation by providing various services, there are also challenges in EVs' charging and discharging management. One of the main challenges is the degradation of the EV battery due to vehicle-to-grid services. Frequent charging and discharging to provide services leads to battery degradation and high costs to EV owners. Also, another challenge is how to encourage EV owners to participate in providing services to the network. EV owners may be unwilling to offer their batteries to the grid operator or EVs aggregator to provide service due to concerns about the SOC requirement for the next trip. Therefore, EV owners should be encouraged to participate in providing services by using appropriate financial incentives. Other challenges include the lack of market frameworks for some services. For example, there is no market framework for reactive power support services. In addition, controlling the reactive power of EVs at the connection point to the network requires special chargers that are not currently widely available. Therefore, the implementation of reactive power support services in real applications faces difficulties. However, in the future, EV owners are expected to be encouraged to participate in the various services by improving charging infrastructure, decreasing the price of EV's batteries, and creating the market framework for different services [5]. According to the International Energy Agency, the whole number of EVs worldwide will catch 130 million by 2030 [3]. Therefore, with rising the number of EVs, smart charging can improve network operation.

6.6 Conclusion

This chapter examined the optimal management of charging and discharging of EVs in the power system to provide services to the power system. Accordingly, EV services available to the power grid were categorized into three classes: active power support services, reactive power support, and support for integrating renewable energy sources into the grid. Frequency control, load variance minimization, loads restoration, loss minimization, and voltage control services were in the class of active power support services, and reactive power loss minimization services and reactive power voltage control services were in the category of reactive power support services. Also, renewable energy integration support services include wind and solar integration support services. Each of these services was examined in detail, and different EV charging and discharging strategies were evaluated to provide these services. According to the reviewed papers, by managing the charge and discharge of EVs, various services can be provided to the network, and the network can be effectively supported. Incentives can also be provided to encourage EV owners to participate in service delivery to the network. In this chapter, the ability to combine different EV services to better support the network was examined. As mentioned, in planning issues for the day ahead, different services can be combined, but for real-time applications, combining some services faces major problems. Then, some of the necessary constraints in the EVs' charging and discharging optimization problem were considered. Mathematical models and solving methods of the optimization problem were stated, and the efficiency of different solving methods was evaluated. Finally, the current status, challenges, and outlook of smart charging of EVs in the power system were presented. Since the ability of EVs to participate in the loads restoration and increase network resilience has received less attention, this can be considered for future research. Also, considering EVs' temporal and spatial uncertainties in optimization problems makes it possible to have more precise management of their charge and discharge.

References

1. Saldaña, G., Martin, J. I. S., Zamora, I., Asensio, F. J., & Oñederra, O. (2019). Electric vehicle into the grid: Charging methodologies aimed at providing ancillary services considering battery degradation. *Energies, 12*(12), 2443.
2. Solanke, T. U., Ramachandaramurthy, V. K., Yong, J. Y., Pasupuleti, J., Kasinathan, P., & Rajagopalan, A. (2020). A review of strategic charging–discharging control of grid-connected electric vehicles. *Journal of Energy Storage, 28*, 101193.
3. Zheng, Y., Niu, S., Shang, Y., Shao, Z., & Jian, L. (2019). Integrating plug-in electric vehicles into power grids: A comprehensive review on power interaction mode, scheduling methodology and mathematical foundation. *Renewable and Sustainable Energy Reviews, 112*, 424–439.
4. Nour, M., Chaves-Ávila, J. P., Magdy, G., & Sánchez-Miralles, Á. (2020). Review of positive and negative impacts of electric vehicles charging on electric power systems. *Energies, 13*(18), 4675.
5. Arias, N. B., Hashemi, S., Andersen, P. B., Træholt, C., & Romero, R. (2019). Distribution system services provided by electric vehicles: Recent status, challenges, and future prospects. *IEEE Transactions on Intelligent Transportation Systems, 20*(12), 4277–4296.
6. Tan, K. M., Ramachandaramurthy, V. K., & Yong, J. Y. (2016). Integration of electric vehicles in smart grid: A review on vehicle to grid technologies and optimization techniques. *Renewable and Sustainable Energy Reviews, 53*, 720–732.
7. Yong, J. Y., Ramachandaramurthy, V. K., Tan, K. M., & Mithulananthan, N. (2015). A review on the state-of-the-art technologies of electric vehicle, its impacts and prospects. *Renewable and Sustainable Energy Reviews, 49*, 365–385.
8. Hu, J., Morais, H., Sousa, T., & Lind, M. (2016). Electric vehicle fleet management in smart grids: A review of services, optimization and control aspects. *Renewable and Sustainable Energy Reviews, 56*, 1207–1226.
9. Tomić, J., & Kempton, W. (2007). Using fleets of electric-drive vehicles for grid support. *Journal of Power Sources, 168*(2), 459–468.
10. Zhang, Q., Li, Y., Li, C., & Li, C. (2018). Grid frequency regulation strategy considering individual driving demand of electric vehicle. *Electric Power Systems Research, 163*, 38–48.

11. Wang, Y., John, T., & Xiong, B. (2019). A two-level coordinated voltage control scheme of electric vehicle chargers in low-voltage distribution networks. *Electric Power Systems Research, 168*, 218–227.
12. Einaddin, A. H., & Yazdankhah, A. S. (2020). A novel approach for multi-objective optimal scheduling of large-scale EV fleets in a smart distribution grid considering realistic and stochastic modeling framework. *International Journal of Electrical Power & Energy Systems, 117*, 105617.
13. Jian, L., Zheng, Y., & Shao, Z. (2017). High efficient valley-filling strategy for centralized coordinated charging of large-scale electric vehicles. *Applied Energy, 186*, 46–55.
14. Mehta, R., Verma, P., Srinivasan, D., & Yang, J. (2019). Double-layered intelligent energy management for optimal integration of plug-in electric vehicles into distribution systems. *Applied Energy, 233*, 146–155.
15. Lili Gong, W., Cao, K. L., Yue, Y., & Zhao, J. (2020). Demand responsive charging strategy of electric vehicles to mitigate the volatility of renewable energy sources. *Renewable Energy, 156*, 665–676.
16. Shahab, M., Wang, S., & Junejo, A. K. (2021). Improved control strategy for three-phase microgrid management with electric vehicles using multi objective optimization algorithm. *Energies, 14*(4), 1146.
17. Peng, C., Zou, J., & Lian, L. (2017). Dispatching strategies of electric vehicles participating in frequency regulation on power grid: A review. *Renewable and Sustainable Energy Reviews, 68*, 147–152.
18. Wu, F., & Sioshansi, R. (2019). A stochastic operational model for controlling electric vehicle charging to provide frequency regulation. *Transportation Research Part D: Transport and Environment, 67*, 475–490.
19. Clairand, J.-M. (2020). Participation of electric vehicle aggregators in ancillary services considering users' preferences. *Sustainability, 12*(1), 8.
20. Liu, H., Huang, K., Wang, N., Qi, J., Qiuwei, W., Ma, S., & Li, C. (2019). Optimal dispatch for participation of electric vehicles in frequency regulation based on area control error and area regulation requirement. *Applied Energy, 240*, 46–55.
21. Amamra, S.-A., & Marco, J. (2019). Vehicle-to-Grid aggregator to support power grid and reduce electric vehicle charging cost. *IEEE Access, 7*, 178528–178538.
22. Wang, L., & Chen, B. (2019). Dual-level consensus-based frequency regulation using vehicle-to-grid service. *Electric Power Systems Research, 167*, 261–276.
23. Fu, H., Han, Y., Wang, J., & Zhao, Q. (2018). A novel optimization of plug-in electric vehicles charging and discharging behaviors in electrical distribution grid. *Journal of Electrical and Computer Engineering, 2018*.
24. Tan, K. M., Ramachandaramurthy, V. K., Yong, J. Y., Padmanaban, S., Mihet-Popa, L., & Blaabjerg, F. (2017). Minimization of load variance in power grids—investigation on optimal vehicle-to-grid scheduling. *Energies, 10*(11), 1880.
25. García-Villalobos, J., Zamora, I., Knezović, K., & Marinelli, M. (2016). Multi-objective optimization control of plug-in electric vehicles in low voltage distribution networks. *Applied Energy, 180*, 155–168.
26. Yang, J., He, L., & Siyao, F. (2014). An improved PSO-based charging strategy of electric vehicles in electrical distribution grid. *Applied Energy, 128*, 82–92.
27. Reddy, K., Ramakrishna, S. M., & Vijayakumar, D. (2019). A novel strategy for maximization of plug-in electric vehicle's storage utilization for grid support with consideration of customer flexibility. *Electric Power Systems Research, 170*, 158–175.
28. Khan, S. U., Mehmood, K. K., Haider, Z. M., Rafique, M. K., Khan, M. O., & Kim, C.-H. (2021). Coordination of multiple electric vehicle aggregators for peak shaving and valley filling in distribution feeders. *Energies, 14*(2), 352.
29. Mazumder, M., & Debbarma, S. (2020). EV charging stations with a provision of V2G and voltage support in a distribution network. *IEEE Systems Journal, 15*(1), 662–671.
30. Sun, W., Kadel, N., Alvarez-Fernandez, I., Nejad, R. R., & Golshani, A. (2019). Optimal distribution system restoration using PHEVs. *IET Smart Grid, 2*(1), 42–49.

31. Momen, H., Abessi, A., & Jadid, S. (2020). Using EVs as distributed energy resources for critical load restoration in resilient power distribution systems. *IET Generation, Transmission & Distribution, 14*(18), 3750–3761.
32. Rahman, T., & Zhihua, Q. (2017). The role of electric vehicles for frequency regulation during grid restoration. In *2017 IEEE Power & Energy Society General Meeting* (pp. 1–5). IEEE.
33. Gong, L., Cao, W., Liu, K., Zhao, J., & Li, X. (2018). Spatial and temporal optimization strategy for plug-in electric vehicle charging to mitigate impacts on distribution network. *Energies, 11*(6), 1373.
34. Islam, M. R., Lu, H., Hossain, J., & Li, L. Multiobjective optimization technique for mitigating unbalance and improving voltage considering higher penetration of electric vehicles and distributed generation. *IEEE Systems Journal, 14*(3), 2020, 3676–3686.
35. Jabalameli, N., Xianging, S., & Deilami, S. (2019). An online coordinated charging/discharging strategy of plug-in electric vehicles in unbalanced active distribution networks with ancillary reactive service in the energy market. *Energies, 12*(7), 1350.
36. Moeini-Aghtaie, M., Abbaspour, A., & Fotuhi-Firuzabad, M. (2014). Online multicriteria framework for charging management of PHEVs. *IEEE Transactions on Vehicular Technology, 63*(7), 3028–3037.
37. Chukwu, U. C., & Mahajan, S. M. (2019). The prospects of V2G for reactive power compensation in electric distribution networks. In *2019 IEEE Power & Energy Society General Meeting (PESGM)* (pp. 1–5). IEEE.
38. Zhang, W., Das, P., & Srinivasan, D. (2016). A vehicle-to-grid based reactive power dispatch approach using particle swarm optimization. In *2016 IEEE Congress on Evolutionary Computation (CEC)* (pp. 4413–4420). IEEE.
39. Madahi, S. S., Karimi, H. N., Abyaneh, H. A., & Marzband, M. (2020). Co-optimization of energy losses and transformer operating costs based on smart charging algorithm for plug-in electric vehicle parking lots. *IEEE Transactions on Transportation Electrification, 7*(2), 527–541.
40. Akhavan-Rezai, E., Shaaban, M. F., El-Saadany, E. F., & Karray, F. (2017). Managing demand for plug-in electric vehicles in unbalanced LV systems with photovoltaics. *IEEE Transactions on Industrial Informatics, 13*(3), 1057–1067.
41. Farahani, H. F. (2017). Improving voltage unbalance of low-voltage distribution networks using plug-in electric vehicles. *Journal of Cleaner Production, 148*, 336–346.
42. Pirouzi, S., Latify, M. A., & Reza Yousefi, G. (2020). Conjugate active and reactive power management in a smart distribution network through electric vehicles: A mixed integer-linear programming model. *Sustainable Energy, Grids and Networks, 22*, 100344.
43. Zeraati, M., Golshan, M. E. H., & Guerrero, J. M. (2017). A consensus-based cooperative control of PEV battery and PV active power curtailment for voltage regulation in distribution networks. *IEEE Transactions on Smart Grid, 10*(1), 670–680.
44. Kikusato, H., Fujimoto, Y., Hanada, S.-i., Isogawa, D., Yoshizawa, S., Ohashi, H., & Hayashi, Y. (2019). Electric vehicle charging management using auction mechanism for reducing PV curtailment in distribution systems. *IEEE Transactions on Sustainable Energy, 11*(3), 1394–1403.
45. Seddig, K., Jochem, P., & Fichtner, W. (2017). Integrating renewable energy sources by electric vehicle fleets under uncertainty. *Energy, 141*, 2145–2153.
46. Li, Y., Ni, Z., Zhao, T., Minghui, Y., Liu, Y., Lei, W., & Zhao, Y. (2020). Coordinated scheduling for improving uncertain wind power adsorption in electric vehicles—wind integrated power systems by multiobjective optimization approach. *IEEE Transactions on Industry Applications, 56*(3), 2238–2250.
47. Huang, Q., Jia, Q.-S., & Guan, X. (2016). Robust scheduling of EV charging load with uncertain wind power integration. *IEEE Transactions on Smart Grid, 9*(2), 1043–1054.
48. Southwest Research Institute. (2014). *SwRI develops first ERCOT-Qualified Vehicle-to-Grid aggregation system*. https://www.swri.org/press-release/swri-develops-first-ercot-qualified-vehicle-grid-aggregation-system

49. Mitsubishi. MitsubishI motors implements first 'vehicle to grid' pilot on Dutch market with Mitsubishi Outland PHEV. http://www.mitsubishi-motors.com/en/newsrelease/2017/detail1082.html
50. Pearre, N. S., & Ribberink, H. (2019). Review of research on V2X technologies, strategies, and operations. *Renewable and Sustainable Energy Reviews, 105*, 61–70.

Chapter 7
Smart Charging of EVs to Harvest Flexibility for PVs

Abbas Rabiee, Andrew Keane, and Alireza Soroudi

7.1 Status Quo, Challenges and Outlook

The owners of electric vehicles (EVs) tend to charge their vehicles when required. This might seem to be normal, however, if all EV owners decide to do so simultaneously it will create serious technical challenges for the distribution networks. One main challenge could be the need for extra investment to reinforce the network. The integration of distributed energy resources (DERs) has created new challenges and opportunities for the distribution network operators. The DERs are from different technologies including EVs, photo-voltaic (PV) cells, batteries, electrolyzer for green hydrogen extraction [1, 2], etc. Hence, the focus of this chapter is on EV and PV integration in distribution networks. The aim is to modify the exiting aggregated EV demand profiles of both residential and public EV chargers in such a way that the available network assets are utilized more efficiently and hence, the need for extra expansion and reinforcement of the grid is minimized. The obtained results imply that for the modified EVs' behavior, the existing network's capability to supply more EV demand is increased significantly as well as the impact of EVs on peak demand increase of the network is minimized. Additionally, more PV energy can be injected along with the increase of EVs penetration.

A. Rabiee (✉) · A. Keane · A. Soroudi
School of Electrical and Electronic Engineering, University College Dublin, Dublin, Ireland
e-mail: abbas.rabiee@ucd.ie; andrew.keane@ucd.ie; alireza.soroudi@ucd.ie

V. Vahidinasab, B. Mohammadi-Ivatloo (eds.), *Electric Vehicle Integration via Smart Charging*, Green Energy and Technology,
https://doi.org/10.1007/978-3-031-05909-4_7

7.2 Introduction

7.2.1 Background and Literature Review

The interest in maximizing the distribution networks' hosting capacity (HC) for EVs is growing globally due to the energy policy action needed to ensure climate neutrality, i.e., net-zero greenhouse gas emissions. The HC is defined as the amount of EV demand that can be connected to the power grid without endangering the reliability of the network and power quality for other customers. In all future energy scenarios worldwide [3–5], there is a step change in the uptake of the electrified transport sector. As an example, in *coordinated action* and *centralized energy* scenarios in Ireland, high degrees of transportation electrification is predicted for the next 20 years [6]. In Ireland, the passenger car fleet increases by 57% (from 2011 with 2.9 million vehicles on the road) by 2050. Moreover, in 2020 the EV contribution to the passenger car segment is 10% which will grow to 60% by 2050 [7]. From the perception of electric utilities, such EV units can play a vital role in enabling the next phase of the energy transition toward high penetrations of sustainable energy sources. The charge/discharge event of EVs can be managed flexibly and efficiently which may provide various regulatory benefits to electricity networks [8, 9].

A comprehensive literature review performed by [10], where the main challenges raised by EVs in power systems are discussed. These challenges are mainly regarding the power quality issues, economy perspectives, reliability, the grid's control, and loadability. In [11], different charging strategies are used to evaluate the reliability of the power system. In [12], an algorithm is developed for microgrid planning with EV charging demand. The aim is to find the most economical configuration to maximize renewable generation utilization. In [13], an optimization model is proposed based on Genetic Algorithm to coordinate the charging of EVs. The technical impacts of large-scale EVs integration on power systems are studied in [14]. In [15], the EVs charge and discharge are jointly managed with other available resources, considering two different objectives, namely the minimization of operation costs and the optimization of the power demand curve. A multi-objective optimization model is developed in [16] by considering three aims, namely: a demand-side management method for managing the operation of a dynamic road system, a regulator that acts to reduce the amount of power drawn from the grid when generation is strained, and a local renewable energy generation and its associated reductions in CO2 emissions.

Shifting the EVs charging to off-peak periods is a key challenge for integration of EVs into the electricity grid on a large scale, that is the subject line of several researches. In [17] consumption tariffs are designed to displace the electric vehicle charging to periods with lower network utility and major PV injections. Also, a methodology for optimization of power systems demand at the presence of EVs is developed in [18], via introducing a statistical model for prediction of the EV charging demand. An EV demand modification model is proposed in [19] by including the network constraints such as transformer and line limitations. Moreover, an approach is proposed in [20] for optimal charging of EVs at the presence of PVs, taking into account the transmission and distribution networks interactions.

7.2.2 Contributions

When more blocks of EVs demand entered to the network, the overall electricity demand will significantly increase because EVs consume a large amount of electricity. Though, if the charging pattern of EVs is locally controlled to satisfy the pre-defined charging plan only, serious technical problems can be observed in distribution systems, such as reverse power flows, voltage violations, and line congestion due to the overloaded feeders [21]. Also, large blocks of EVs charging demand can increase the peak demand of the electricity network and consequently imposes high amount of grid reinforcement costs. Hence, optimization of EVs demand profile is vital both economically and technically. In order to avoid the aforementioned problems, as well as utilize the benefits of EVs, it is necessary to design a time-of-use (TOU) tariff for EVs' demand. By proper design of TOU tariffs, the network operator can be assured of maximum HC of the network, as well as minimum need for grid upgrade. This chapter proposes a model for modification of EVs' charging profile from which the policymakers can design proper TOU tariffs.

The following factors are considered in the proposed approach: the hourly demand profile of the network, the existing EV chargers' (EVCs) rating as well as the existing EV demand behavior in residential and public EVCs. The aim is to determine the optimal EV charge pattern considering the technical constraints of the network. The graphical abstract of this chapter is shown in Fig. 7.1. Also, the future energy transition scenarios worldwide substantiate the key role of renewable energy sources (RESs) in the next decades. In distribution levels, the prominent type of RESs is PV, because of its easy installation and scalability. Hence, in this chapter, the EVs demand profile is modified by considering the PVs power generation. The following cases are studied in this chapter.

- Distribution networks HC maximization for EVs, without considering any PVs injection.
- Distribution networks HC maximization for EVs, with considering aggregated PVs.

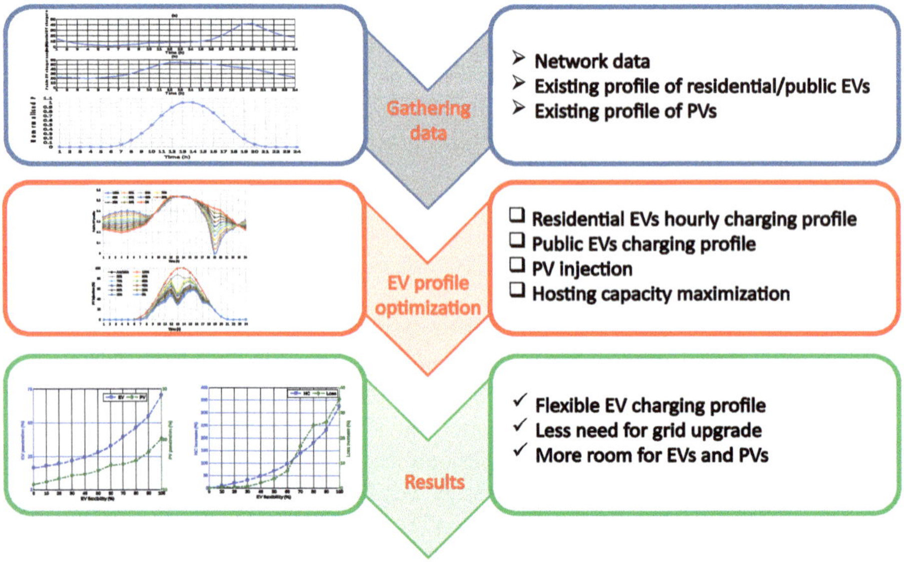

Fig. 7.1 Graphical abstract of the proposed methodology for smart charging profile determination

7.2.3 Chapter Organization

The rest of this chapter is organized as follows: Sect. 7.3 describes the proposed optimization model for the modification of EVs charging pattern. The case studies and numerical results are given in Sect. 7.4. Finally, Sect. 7.5 concludes this chapter.

7.3 Determination of Optimal EV Demand Profile

7.3.1 Assumptions

By growing the number of EVs, their inherent impacts on the electricity grid as well as electricity markets will be increased. The most important influencing factors are the quantity and types of EVs, EVs energy usage, types and locations of EVCs, and the charging behavior of vehicle owners. Vehicle charger technology has the potential to minimize the potential impact of EVs demand on the electricity system as well as the electricity markets. The EVCs' technology is evolving from simple chargers and patterns that are readily available today, to smart chargers with features such as programmable charging start times to smarter charging technology that optimizes vehicle charging in line with dynamic electricity price signals. As an example, [22] requires that new private charge points sold in UK must be smart and meet certain device-level requirements. In the highly EV-penetrated distribution systems, obtaining the optimized EVs' charging profile is necessary in order to ensure the minimum need for grid development as well as additional generation capacity. In this section, an optimization model is proposed to determine the optimal EV charging pattern in an hourly basis. The proposed model has the following features:

- Optimal hourly charge patterns are determined for both residential and public EVCs.
- The existing demand pattern of EVs is considered as the baseline, and the EVs' demand pattern modified by definition of a flexibility factor. This flexibility factor gives insight for policymakers to change the existing behavior of EVs to more optimized patterns.
- The demand pattern of EVs is optimized to maximize the EV energy penetration.
- The MV distribution network technical constraints such as power flow, current, and voltage limits are satisfied in the entire horizon.

7.3.2 Mathematical Formulation

Objective Function
The aim is to maximize the HC distribution network for EVs energy demand , as follows:

$$\max \quad HC = \sum_{b,t,e} PE_{b,t}^{e} \tag{7.1}$$

$$\textit{Subject to} \ :$$

$$(7.2) \ \textit{to} \ (7.15)$$

where HC is the distribution network HC for EVs, e is the index for EVC type (residential or public), b is the index for electrical network buses, and t is the index for operation intervals. Besides, $PE_{b,t}^{e}$ is the hourly demand of the e-th type of EVs at time t and bus b.

Problem Constraints

- Electricity network constraints:
  The full AC power flow constraints are considered as follows:

$$PG_{b,t} + PV_{b,t} - PD_{b,t} - \sum_{e \in \Omega_{EV}} PE_{b,t}^{e} = \sum_{k \in \Omega_{B_b}} P_{bk,t} \tag{7.2}$$

$$QG_{b,t} - QD_{b,t} = \sum_{k \in \Omega_{B_b}} Q_{bk} \tag{7.3}$$

$$P_{bk,t} = +g_{bk} V_{b,t}^2 - V_{b,t} V_{k,t} \left(g_{bk} cos\left(\theta_{bk,t}\right) + b_{bk} sin\left(\theta_{bk,t}\right) \right) \tag{7.4}$$

$$Q_{bk} = -b_{bk} V_{b,t}^2 - V_{b,t} V_{k,t} \left(g_{bk} sin\left(\theta_{bk,t}\right) - b_{bk} cos\left(\theta_{bk,t}\right) \right) \tag{7.5}$$

$$\left(P_{bk,t}\right)^2 + \left(Q_{bk,t}\right)^2 = \left(V_{b,t} I_{bk,t}\right)^2 \tag{7.6}$$

$$0 \le I_{bk,t} \le \overline{I}_{bk} \qquad \forall b \in \Omega_B, \forall k \in \Omega_{B_b} \tag{7.7}$$

$$\underline{PG}_b \le PG_{b,t} \le \overline{PG}_b \qquad \forall b \in \Omega_{B_S} \tag{7.8}$$

$$\underline{QG}_{b,t} \le QG_{b,t} \le \overline{QG}_{b,t} \qquad \forall b \in \Omega_{B_S} \tag{7.9}$$

$$\underline{V}_b \le V_{b,t} \le \overline{V}_b \qquad \forall b \in \Omega_B \tag{7.10}$$

$$0 \le PV_{b,t} \le \Gamma_t \times CPV_b \qquad \forall b \in \Omega_{B_{PV}} \tag{7.11}$$

where (7.2) and (7.3) are the active and reactive power balances. Also, (7.4)–(7.6) are the active, reactive, and current flow of the line connecting the buses b and k. Besides, the feeder's current flow limit is modeled by (7.7). Also, (7.8) and (7.9) are the upper/lower limits of active/reactive power exchange with the upstream network, and (7.10) is the voltage limits, respectively. Besides, the hourly PV injection is expressed by (7.11). The variables and sets used in the above formulations are defined as follows: $(PG/QG)_{b,t}$ is the active/reactive power generation in node b at time t. $(P/Q)_{bk,t}$ is the active/reactive power flowing through bk-th line at time t. $I_{bk,t}$ is the current flowing through the bk-th line at time t. $(g/b)_{bk}$ is the conductance/susceptance of the bk-th line. $(V/\theta)_{b,t}$ is the voltage magnitude/angle of bus b at time t. β is the flexibility factor of EV demand profile. $(PD/QD)_{b,t}$ is the active/reactive demand in bus b at time t. $(\overline{PG}/\underline{PG})_{b,t}$ is the upper/lower limit of active power generation in bus b and time t. $(\overline{QG}/\underline{QG})_{b,t}$ is the upper/lower limit of reactive power generation in bus b and time t. $(\overline{V}/\underline{V})_b$ is the upper/lower limit of voltage magnitude at bus b. $\overline{I}_{bk}$ is the upper limit of the current flowing through bk-th line. $PV_{b,t}$ is the active power injection by PVs at bus b and time t. Γ_t is the available percentage of PV at time t. CPV_b is the nominal capacity of PV connected to bus b. Also, Ω_B is the set of network buses. Ω_{B_S} is the set of network buses connected to the upstream network. Ω_{B_b} is the set of network buses connected to bus b. Ω_{EV} is the set of EVC types (residential or public). $\Omega_{B_{PV}}$ is the set of network buses with PV units.

- EVs' demand profile flexibility:

The current charging profile of EVCs is modified by the proposed model in a way that as much as possible, more EVs demand can be supplied by the network. Hence, a flexibility factor, (i.e., $0 \le \beta \le 1$), is defined as follows to change the existing profile $\hat{\Xi}_t^e$ to the modified profile of Ξ_t^e ($\forall t \in \Omega_T, \forall e \in \Omega_{EV}$ and $\forall b \in \Omega_{B_{EV}}$).

$$\varepsilon_t^e = \hat{\Xi}_t^e + \beta\left(\Xi_t^e - \hat{\Xi}_t^e\right) \tag{7.12}$$

$$0 \leq PE_{b,t}^e \leq \varepsilon_t^e \times CEV_b^e \tag{7.13}$$

$$\sum_t \varepsilon_t^e = \sum_t \hat{\Xi}_t^e \tag{7.14}$$

$$0 \leq \varepsilon_t^e \leq 1 \tag{7.15}$$

where the modified EVCs' demand profile is calculated by (7.12) and EVCs' active power demand is stated as a function of modified profile by (7.13). Also, (7.14) ensures that the energy demand of EVCs remains the same in the entire operation horizon and the modified charging profile of EVCs is limited by (7.15). Besides, ε_t^e is the modified demand profile of EV type e at time t. $\hat{\Xi}_t^e$ is the existing demand profile of EV type e at time t. Ξ_t^e is the fully optimized demand profile of EV type e at time t. CEV_b is the nominal capacity of EVC connected to bus b. $\Omega_{B_{EV}}$ is the set of network buses with EVCs and Ω_T is the set of time intervals.

In order to determine the optimized EVs' profiles based on the above model, one should determine Ξ_t^e in (7.12), firstly. Hence by setting $\beta = 1$, the optimal profile, i.e., Ξ_t^e, could be attained. Then, by keeping in the hand both Ξ_t^e and $\hat{\Xi}_t^e$, one can determine the modified profiles of EVs for any level of flexibility factor.

7.4 Numerical Studies

The proposed non-linear model is implemented in GAMS [23] and solved using CONOPT4 [24] solver.

7.4.1 Data

To obtain the optimal demand pattern of EVs, case studies are carried out on the IEEE 69-bus MV radial distribution feeder in general algebraic modeling system (GAMS), which is a standard package for dealing with optimization problems [23]. The one-line representation of this system is given in Fig. 7.2. The network data is adopted from [25]. The current limit of the feeder (in all sections) is assumed to be $270A$, whereas the limits on the active and reactive power absorbed from the substation that is connected to the upstream HV network are 6 MW and 6 MVAr, respectively. Also the voltage upper/lower limit is $1.100/0.900$ pu. The hourly demand profile of this network, in terms of active and reactive powers, is shown in Fig. 7.3. It is observed that the critical demand interval is $(10, 22)$ where the peak demand occurs in hour 19.

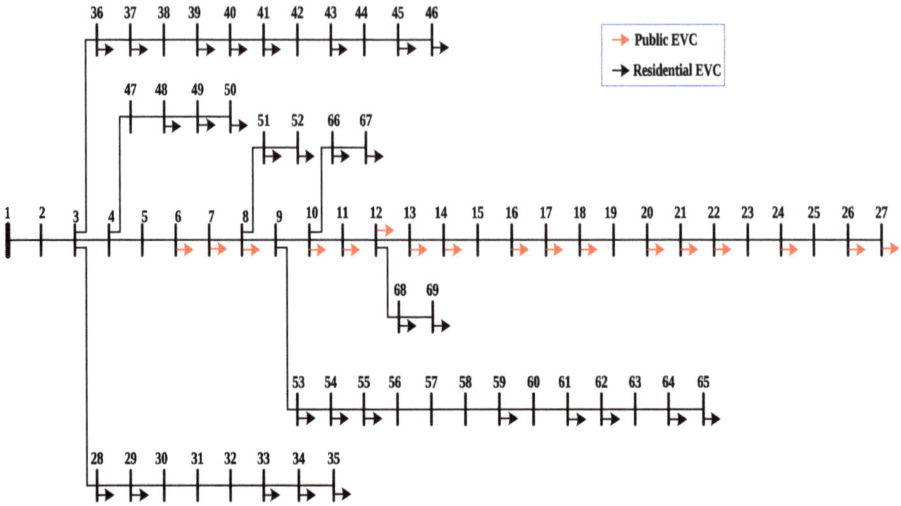

Fig. 7.2 One-line diagram of the IEEE 69-bus radial distribution system

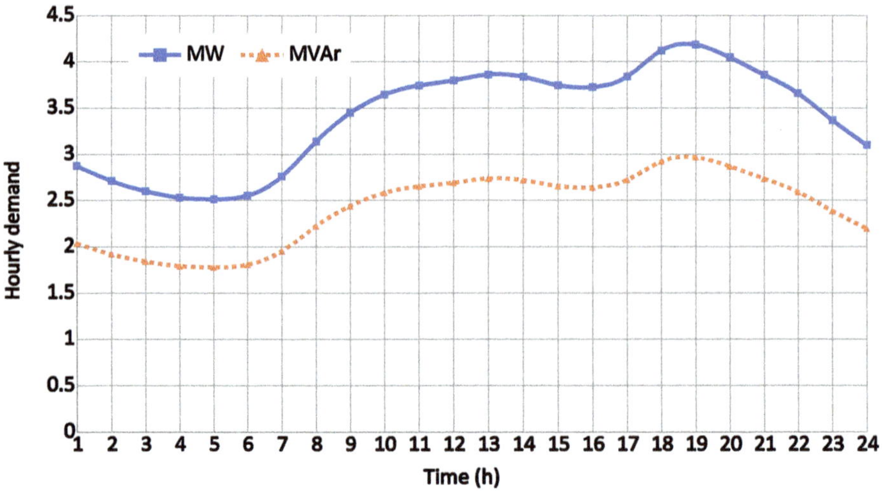

Fig. 7.3 Hourly active and reactive power demands

The aim is to optimize the EVs' charging demand, taking into account their existing charging behavior for both residential and public chargers. It is assumed that the residential EVCs type is *Standard Type-2 (ST2)*, while the fast chargers (*AC43, CCS, and CHAdeMO*) are commonly employed in public locations because of their fast charging capabilities. Moreover, it is assumed that the public EVCs are located on the main feeder (backbone feeder as shown in Fig. 7.2), each charge point with the aggregated capacity of $4 \times 50\,\text{kW}$. Additionally, the residential EVCs are installed in the spurs, each load point with an aggregated capacity of $10 \times 22\,\text{kW}$. It

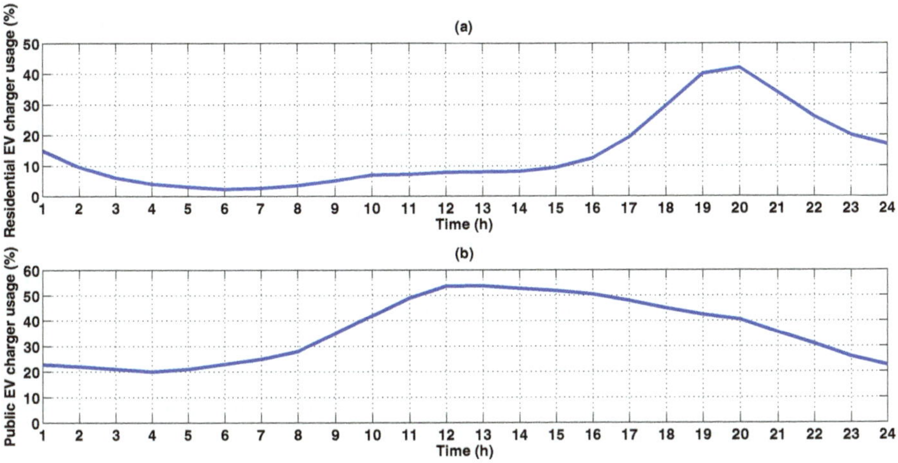

Fig. 7.4 Existing EV charging patterns, (**a**): residential EVCs [26], (**b**): public EVCs [27]

is worth to mention that the residential chargers are basically single phase and in the range of 3.5 − 7.0 kW. But, as the proposed model is basically a generic optimization model and a balanced three-phase MV distribution network is modeled, inclusion of single-phase EVC is not possible directly, except that multiple 3 single-phase EVC are considered. If it is assumed that each single-phase residential EVC rating is about 7.0 kW, it is equivalent with a 21 kW three-phase EVC, as already considered in this study.

The existing charging demand behavior for both residential and public EVCs are depicted in Fig. 7.4. These profiles are adopted from [26] and [27], respectively. These profiles which have been obtained via a large data set of realistic EV charge events in the interval 2017–2019 in UK and Ireland reveal the mean daily utilization of each type of EVC. It is evident from Fig. 7.4a that for the initial profiles, the residential EVs' demand is mainly existing during the early night interval, which coincides with the peak demand of the network. Contrarily, as shown by Fig. 7.4b, the public EVCs are mainly used during the daytime interval. Based on the demand profile given in Fig. 7.3, these types of EV demand behavior impose more pressure on the network since the added EV demand increases the system's peak demand.

As it is aforementioned, to show the mutual impact of EVs and PVs, two cases are considered as follows:

- Case-I: EVs profile optimization, without considering PVs and demand uncertainty.
- Case-II: EV profile optimization, taking into account the PVs generation and without considering demand uncertainty.

7.4.2 Case-I: EVs Profile Optimization, Without Considering PVs

In this case, the aim is to modify the EVs demand profile from the existing pattern given in Fig. 7.4 to an optimal pattern, keeping the following considerations:

- The energy demand of EVs does not change. In other words, each EVC station absorbs constant energy from the network at the entire horizon.
- The EV demand profile should be modified in a way that the highest utilization is made from the existing assets, without reinforcement or upgrade of the network infrastructure.
- The impact of EVs' demand on the system peak loading (and hence, the minimum requirement to investment on the generation capacity expansion) should be minimized.

By optimization of EVs' demand profile, taking into account the above considerations, the modified demand profiles of EVCs vs the flexibility factor are obtained for both residential and public EVCs. As it is aforementioned, at first by setting $\beta = 1$ in (7.12) the fully modified profile is determined. Then, by varying β, the corresponding modified profiles for each level of flexibility can be obtained. The evolution of residential and public EVCs' profiles for various flexibility factors is shown in Fig. 7.5a, b, respectively. Without any percentage of flexibility, the EV profiles are the corresponding initial ones, depicted in Fig. 7.4. By increasing the flexibility factor, both profiles are modified to meet the aforementioned requirements, namely the EV demand in peak hours is reduced and shifted to the light

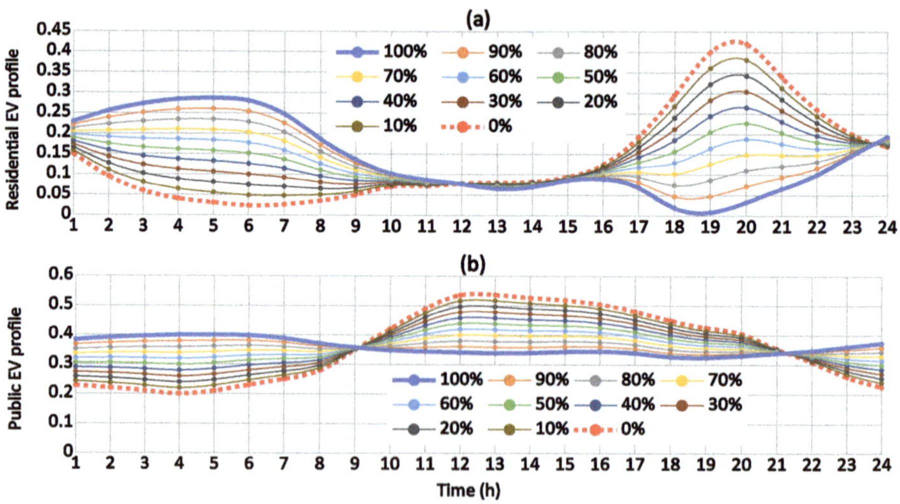

Fig. 7.5 EV profiles evolution versus flexibility degrees *Case-I*: (**a**) Residential EVCs, (**b**) Public EVCs

loading interval (i.e., the early morning interval). More specifically, it is inferred from Fig. 7.5a that in the modified profile of residential EVCs, the EVs demand reduced in the interval (17 22) considerably and displaced to the interval (1 9), which is the light loading interval of the network. During the interval of (10 16) the demand profile of residential EVs does not change considerably. In contrast, for the public EVs, the demand in the interval (10 20) tends to be reduced and shifted to the light loading condition, i.e., the interval (1 8) and (22 24). Hence, from the EV policy makers' viewpoint, the following conclusions can be drawn in this regard:

- For residential EVs, incentive signals should be concentrated to the intervals (17 22) and (1 9), persuading the EV owners to change their charging events from the former interval to the latter.
- For public EVs, incentive programs should be focused on the reduction of charging events in the interval (10 20) and to increase it in the remaining intervals.

The distribution network HC increase for EVs in the studied network, as well as the energy consumption penetration (i.e., the total energy absorbed by EVs divided into the total energy consumption of remaining loads in the entire horizon), are also depicted in Fig. 7.6. It is evident from this figure that by increasing the flexibility from 0% to 100% (i.e., $\beta = 0$ to $\beta = 1$), the HC increases about 225%. This means that in the EV profiles corresponding to higher levels of flexibility, all EVC stations could be used simultaneously. Also, it is observed that the EVs energy consumption penetration is also increased about 175%, which justifies the possibility of feeding

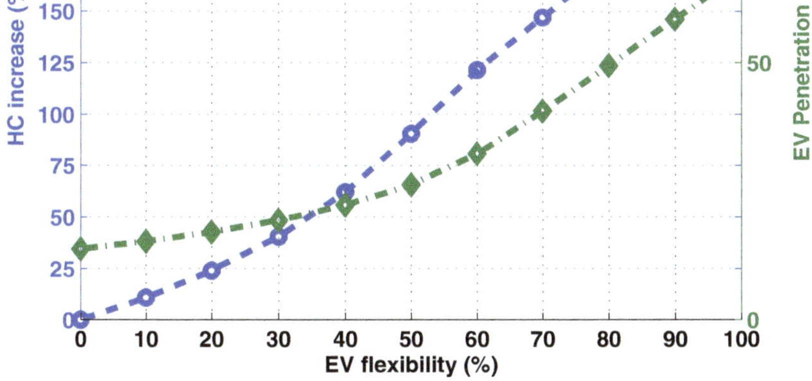

Fig. 7.6 EV penetration level and the HC vs flexibility in *Case-I*

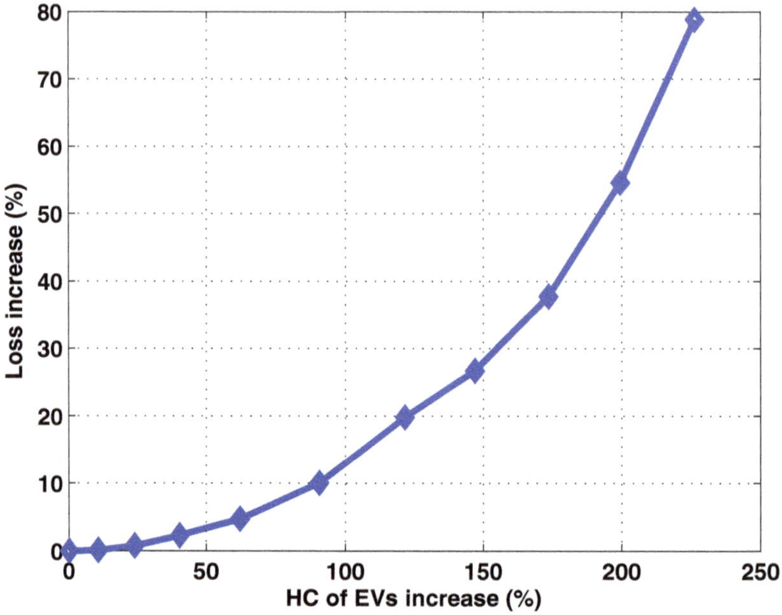

Fig. 7.7 Total energy loss increase vs HC in *Case-I*

more EVs through the existing network (i.e., without any further reinforcement of the network infrastructures). Despite the possibility of supplying more EVs, the energy loss of the network increases, since the current flowing through the branches increases. Figure 7.7 shows the variation of energy loss versus the network's HC. It is observed that for the maximum level of HC which corresponds to a flexibility level of 100%, the energy loss increases about 80%. Also, the energy loss varies more sharply in higher flexibility levels than that of lower flexibilities.

The evolution pattern of the total EV demand profile for various flexibility percentages is depicted in Fig. 7.8. It is observed from this figure that for flexibility levels greater than zero, the EVs do not add any extra demand at the critical interval of peak hours (i.e., the interval (19 20)). This is an important aspect from the economic point of view since adding any demand at peak hours equals to further need for generation capacity expansion.

Finally, the average hourly voltage profile of the network is given in Fig. 7.9. It is inferred from this figure that EV demand profile modification through introducing the flexibility levels has relatively low impact on the voltage profile of network, as can be observed that the voltage difference is less than 0.02 pu for flexibilities 0% and 100%, as well as this difference is less than 0.01 pu in peak demand hour.

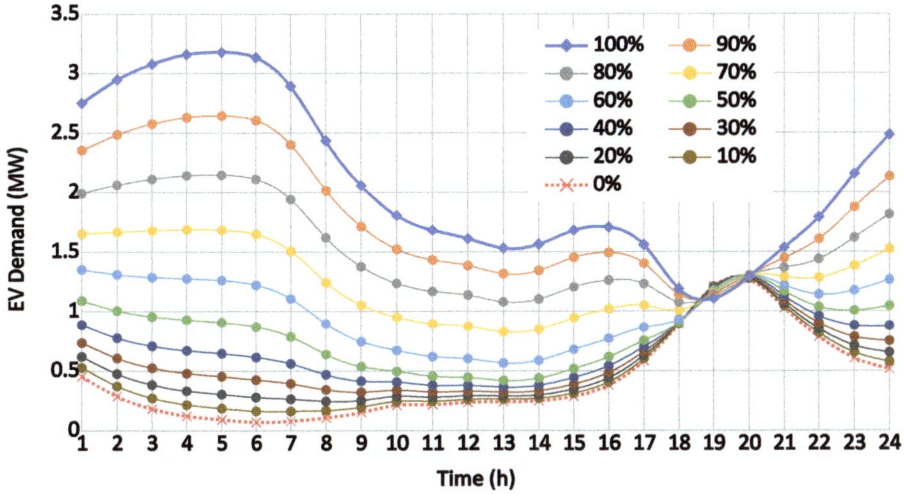

Fig. 7.8 Hourly EV demand profile vs flexibility in *Case-I*

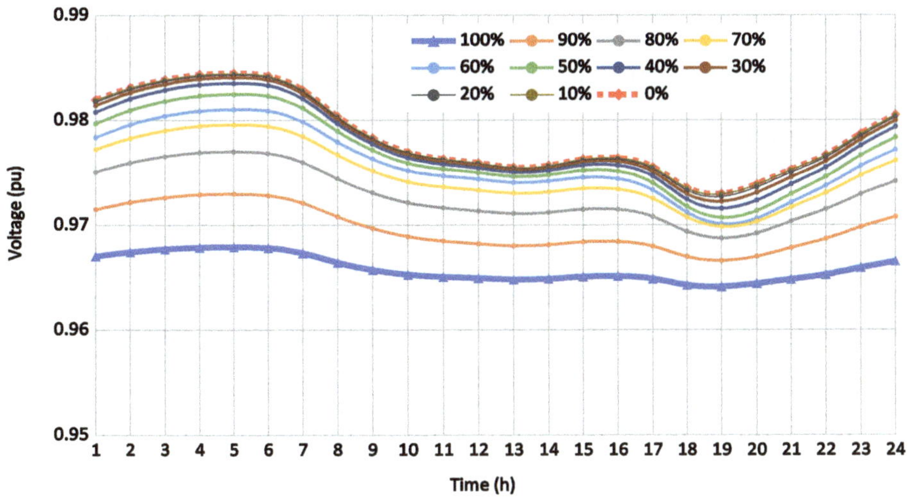

Fig. 7.9 Hourly mean voltage profile vs EV flexibility in *Case-I*

7.4.3 *Case-II: EVs Profile Optimization, Considering PVs*

Integration of RESs such as PVs (as the most common type of RESs in distribution levels) can have a considerable impact on the EVs profile and consequently the network's HC for EV demand. On the other hand, EVs energy consumption can increase the energy injection via PVs, since the operational barriers such as voltage rise and feeder thermal limits can be alleviated. This bi-directional inter-dependency

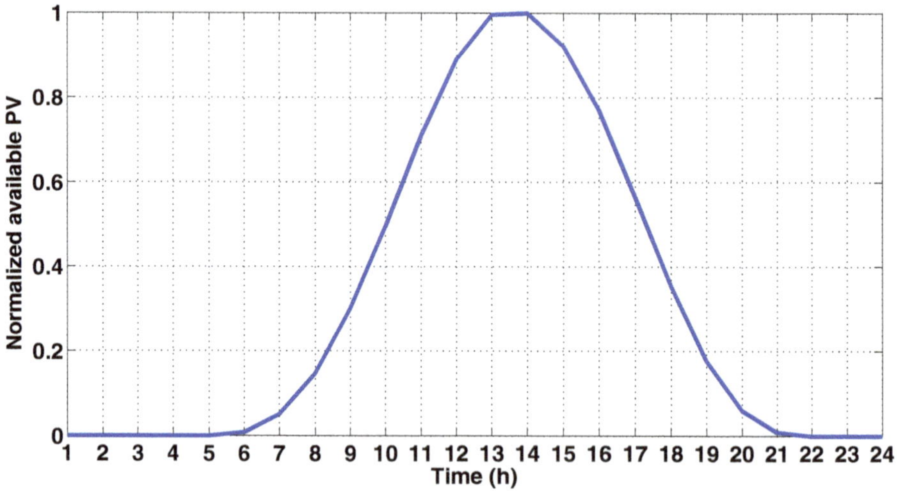

Fig. 7.10 Hourly profile of PVs

between EVs and PVs can provide further flexibility for the energy policymakers to design proper energy tariffs for EVs. In this section, another study is conducted by including the impact of PVs. The PVs data is adopted from the UK database for the interval of 1/1/2020 to 26/6/2020 [28]. The mean PV profile (i.e., Γ_t in (7.11)) that is normalized based on its peak injection is depicted in Fig. 7.10.

In this case, the network is stressed by adding 10% extra demand than that of *Case-I* (i.e., the demand that is given in Fig. 7.3 is multiplied by 1.10). Also the number of EVCs increased by 50% with respect to *Case-I*, in order to investigate the impacts of EV profile modification on the network, especially during the peak hours. Also, it is assumed that $20 \times 2.3\,\text{kW}$ aggregated PV generation capacities are installed at each load point (i.e., the points that the EVCs are also installed in Fig. 7.2).

Figure 7.11 shows the modified EV profiles for different levels of flexibility. In Fig. 7.11a the modified hourly profiles of residential EVCs are depicted.

By comparing this figure and Fig. 7.5a, it is observed that contrary to *Case-I*, because of PVs injection during daytime, more EV demand is supplied during the interval (10 16). It means that in this case a uniform TOU tariff can be defined for the interval (1 16). Also, similar to *Case-I*, the residential EVs' demand during the night hours, i.e., (17 22), should be decreased by imposing a proper TOU tariff. Since the public EVCs are mainly used in daytime, the situation is quite different for these chargers in *Case-II*, in comparison with *Case-I*. It is inferred from Fig. 7.11b that during the interval (10 16)-the busiest hours of the day for public EVCs- the EVs demand profile requires the least modification which means that no incentivizing TOU tariffs is required for this interval in *Case-II*. But, for the interval (17 22) which is also the system peak demand interval, since the PV generation decreases considerably, EVs' demand reduction is necessary; hence, the

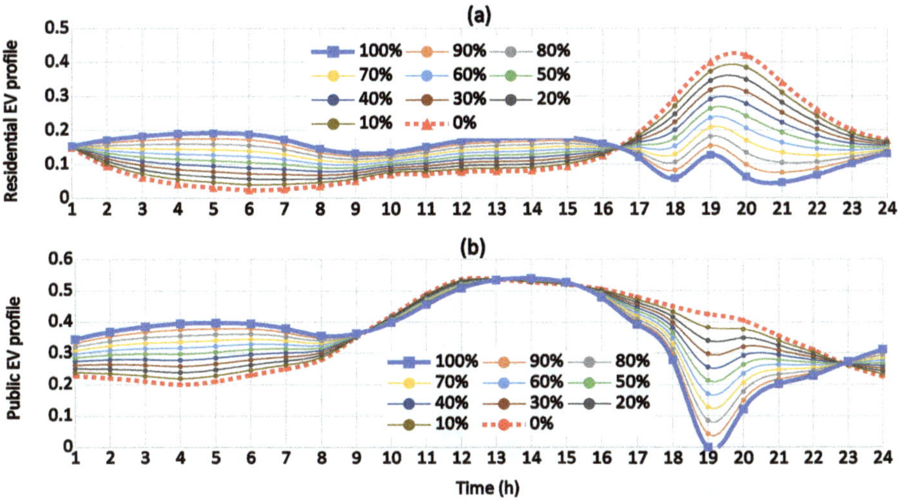

Fig. 7.11 EV profiles versus flexibility degrees in *Case-II*; (**a**): Residential EVCs, (**b**): Public EVCs

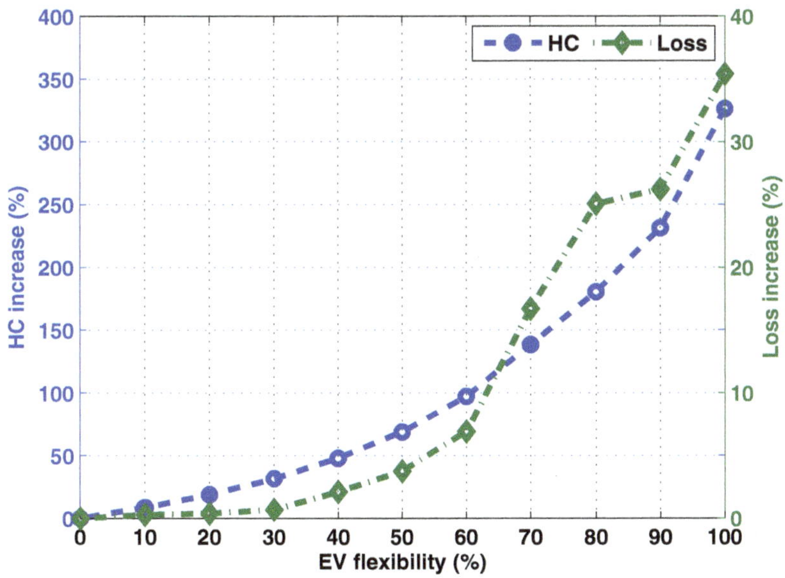

Fig. 7.12 HC and loss increase vs EV flexibility in *Case-II*

TOU tariffs should be designed in a way that the public EVCs demand shifted to the early morning, i.e., (1 8), that is almost similar to *Case-I*. The impact of EVs demand profile modification on both the HC and energy loss is shown by Fig. 7.12. It is evident from this figure that the HC increases vs flexibility such that it is

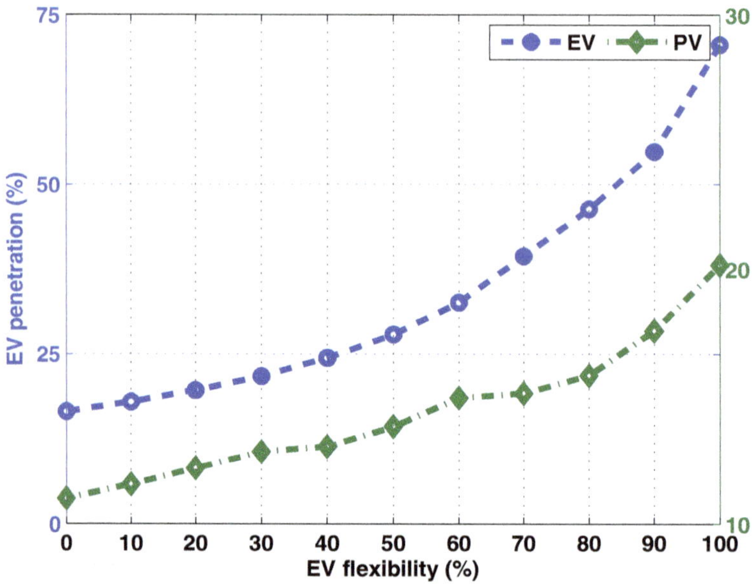

Fig. 7.13 EVs and PVs variation vs EV flexibility in *Case-II*

increased more than threefold. This is much more than that obtained in *Case-I*. Also, the energy loss is increased in this case vs flexibility (about 35% for the flexibility degree of 100%), but its growth is less than that in *Case-I*. This is due to the existence of distributed PV generations that reduces the power procurement from the upstream network (through bus 1). The penetration of EVs and PVs vs the flexibility is depicted in Fig. 7.13. It is observed that the penetrations of EVs and PVs are increased from the initial value of 16% and 11% (corresponding to flexibility level of 0%), to 70% and 20%, respectively, for flexibility of 100%. This means that modification of EV profile can increase the penetration level of both EVs and PVs.

This is shown more clearly in Figs. 7.14 and 7.15. Figure 7.14 reveals the hourly EV demand vs flexibility. It is observed from this figure that the EV demand is shifted to daytime interval and no extra demand added at peak hours.

Also, the PVs power injection vs flexibility is given by Fig. 7.15. It is inferred from this figure that by increasing the flexibility, better utilization of available PV generation capacity is accomplished, such that for flexibility of 100%, the whole PV generation capacity is utilized. Finally, the hourly mean voltage profile of the network in this case is depicted in Fig. 7.16.

The historic data of EV charging is used for finding the optimal charging profile. In order to consider the uncertainties in consumer behavior, PV generation, availability of network components or other input data some appropriate tools [29] should be used.

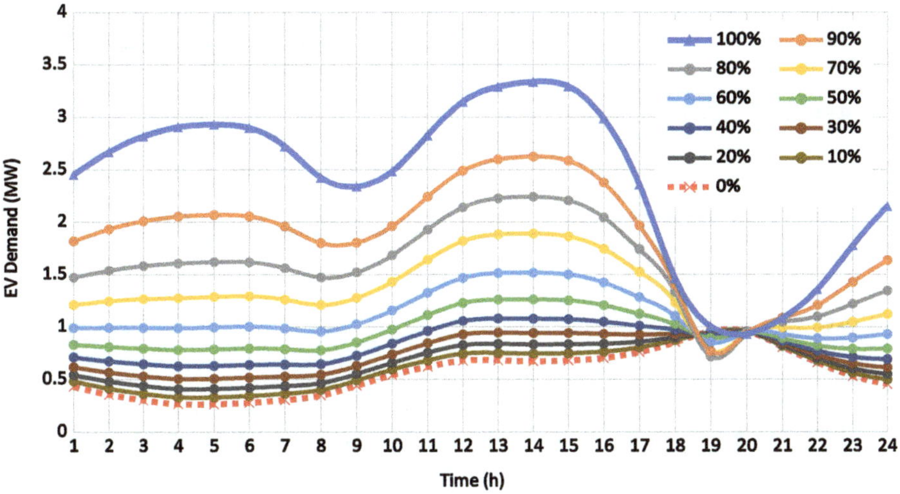

Fig. 7.14 HC and loss increase vs EV flexibility in *Case-II*

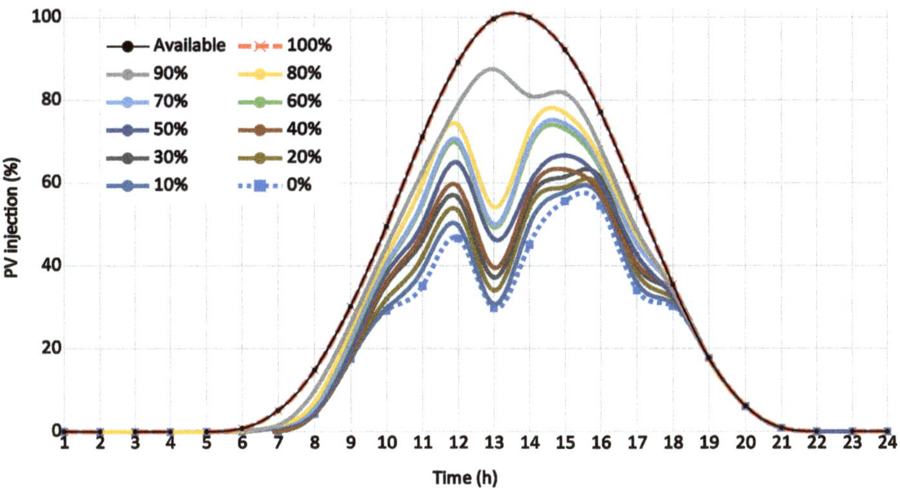

Fig. 7.15 PVs hourly power injections for various EV flexibilities in *Case-II*

7.4.4 Comparative Analysis of Cases

From the energy policymakers perspective, the following conclusions can be drawn based on the performed numerical studies:

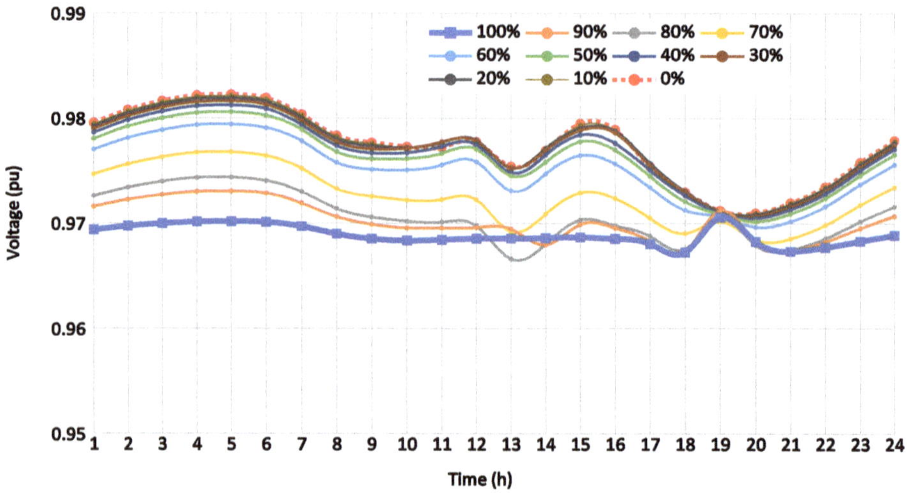

Fig. 7.16 Hourly mean voltage profile vs EV flexibility in *Case-II*

Case-I (without PV generation):

- For residential EVs, the incentivizing programs (such as TOU tariffs) should be made in the intervals (17, 22) and (1 9), such that these programs promote the EV owners to change their charging events from the former interval to the latter.
- For public EVs, incentivizing programs should be focused on the reduction of charging events in the interval (10 20) and to increase it in the remaining intervals.

Case-II (with PV generation, and without demand uncertainty):

- Integration of PVs in distribution networks has an undeniable impact on the EVs profile, which should be considered in TOU tariffs design.
- Particularly, for residential EVs a uniform TOU tariff can be defined for the interval (1 16) at the presence of PVs. Also, during the interval (10 16)- the most interesting interval for public chargers' clients- the EVs demand profile has the least modification at the presence of PVs, meaning that no incentivizing TOU tariffs is required to decrease the EVs' demand at this interval.
- For the interval of (17 22) that coincides with the system peak demand, since the PV generation decreases considerably, demand reduction is necessary for both residential and public EVCs. Consequently, the TOU tariffs should be designed in a way that the EVs demand shifted to the daytime interval, preferably to the early morning hours (i.e., the interval (1 8)).
- The modification of the EV profile by the defined flexibility level can increase the penetration level of both EVs and PVs.

- Increase of energy loss subject to EVs flexibility, in this case, is less than *Case-I*. It proves the role of distributed generations such as PVs in the reduction of the network losses.

7.5 Conclusion

In this chapter, the public and residential EVCs' daily demand profile is modified in a way that the EVs charging events attracted to periods with lower network utility and higher PVs' available power injection. To have a flexible charging profile for EVs, a flexibility factor is introduced. By this modification on EVs charging behavior, the existing assets are utilized more efficiently and hence the need for grid reinforcement and further investments is minimized. Two different cases are studied to highlight the impact of PVs in the smarter charging behavior of EVs. In the first case, the EVs charging profile is modified at the absence of PVs, while in the second case, PVs power injection is considered and the EVs demand profile is obtained by regarding the PVs impact. Also, the modified EVs profile has a minimum impact on the system peak loading increase.

Acknowledgments This book has been partly funded by the Irish Government Department of Communications, Climate Action and Environment. It has also been supported in part by a research Grant from Science Foundation Ireland (SFI) under the SFI Strategic Partnership Programme Grant No. SFI/15/SPP/E3125. The views expressed are those of the authors alone.

Bibliography

1. Rabiee, A., Keane, A., Soroudi, A. (2021). Technical barriers for harnessing the green hydrogen: A power system perspective. *Renewable Energy, 163*, 1580–1587.
2. Rabiee, A., Keane, A., Soroudi, A. (2021). Green hydrogen: A new flexibility source for security constrained scheduling of power systems with renewable energies. *International Journal of Hydrogen Energy, 46*(37), 19270–19284.
3. Global, E. V. (2020). *Outlook 2020*. Paris: IEA.
4. Mintzer, I., Leonard, J. A., Schwartz, P. (2003). *US energy scenarios for the 21st century*. Pew Center on Global Climate Change.
5. Future Energy Scenarios (2019). *UK National Grid Electricity System Operator (ESO)*.
6. *Tomorrow's Energy Scenarios 2019 Ireland*. Ireland: EirGrid (2019).
7. *Electric Vehicle Roadmap*. Ireland: SEAI (2019).
8. Paterakis, N. G., Gibescu, M. (2016). A methodology to generate power profiles of electric vehicle parking lots under different operational strategies. *Applied Energy, 173*, 111–123.
9. Heuberger, C. F., Bains, P. K., Mac Dowell, N. (2020). The evolution of the power system: A spatio-temporal optimisation model to investigate the impact of electric vehicle deployment. *Applied Energy, 257*, 113715.
10. Shaukat, N., Khan, B., Ali, S., Mehmood, C., Khan, J., Farid, U., Majid, M., Anwar, S., Jawad, M., Ullah, Z. (2018). A survey on electric vehicle transportation within smart grid system. *Renewable and Sustainable Energy Reviews, 81*, 1329–1349.

11. Shafiq, S., Irshad, U. B., Al-Muhaini, M., Djokic, S. Z., Akram, U. (2020). Reliability evaluation of composite power systems: Evaluating the impact of full and plug-in hybrid electric vehicles. *IEEE Access, 8*, 114305–114314.
12. Yoon, S.-G., Kang, S.-G. (2017). Economic microgrid planning algorithm with electric vehicle charging demands. *Energies, 10*(10), 1487.
13. Alonso, M., Amaris, H., Germain, J. G., Galan, J. M. (2014). Optimal charging scheduling of electric vehicles in smart grids by heuristic algorithms. *Energies, 7*(4), 2449–2475
14. Boudina, R., Wang, J., Benbouzid, M., Khoucha, F., Boudour, M. (2020). Impact evaluation of large scale integration of electric vehicles on power grid. *Frontiers in Energy, 14*(2), 337–346.
15. Morais, H., Sousa, T., Vale, Z., Faria, P. (2014). Evaluation of the electric vehicle impact in the power demand curve in a smart grid environment. *Energy Conversion and Management, 82*, 268–282.
16. Humfrey, H., Sun, H., Jiang, J. (2019). Dynamic charging of electric vehicles integrating renewable energy: a multi-objective optimisation problem. *IET Smart Grid, 2*(2), 250–259.
17. Lagarto, M., Pinto, J. F., Ferreira, L. (2017). Optimisation of low-voltage distribution networks in a strong embedded microgeneration and electric vehicle penetration context. *CIRED-Open Access Proceedings Journal, 2017*(1), 1922–1926.
18. Zhang, P., Qian, K., Zhou, C., Stewart, B. G., Hepburn, D. M. (2012). A methodology for optimization of power systems demand due to electric vehicle charging load. *IEEE Transactions on Power Systems, 27*(3), 1628–1636.
19. De Hoog, J., Alpcan, T., Brazil, M., Thomas, D. A., Mareels, I. (2014). Optimal charging of electric vehicles taking distribution network constraints into account. *IEEE Transactions on Power Systems, 30*(1), 365–375.
20. Rabiee, A., Keane, A., Soroudi, A. (2021). Enhanced transmission and distribution network coordination to host more electric vehicles and *pv. IEEE Systems Journal*, 1–12. https://doi.org/10.1109/JSYST.2021.3092785.
21. Ali, A., Raisz, D., Mahmoud, K. (2019). Voltage fluctuation smoothing in distribution systems with res considering degradation and charging plan of EV batteries. *Electric Power Systems Research, 176*, 105933.
22. Government, U. (2021). https://www.gov.uk/government/consultations/electric-vehicle-smart-charging (Online; accessed 25 November 2021).
23. Soroudi, A. (2017). *Power system optimization modeling in GAMS*. Berlin: Springer.
24. Drud, A. S. (1994). Conopt—a large-scale GRG code. *ORSA Journal on Computing, 6*(2), 207–216.
25. Zimmerman, R. D., Murillo-Sánchez, C. E., Thomas, R. J. (2010). Matpower: Steady-state operations, planning, and analysis tools for power systems research and education. *IEEE Transactions on Power Systems, 26*(1), 12–19.
26. Dodson, T., Slater, S. (2019). *Electric vehicle charging behaviour study: Final report for national grid ESO*. Cambridge : Element Energy Limited.
27. Pallonetto, F., Galvani, M., Torti, A., Vantini, S. (2020). A framework for analysis and expansion of public charging infrastructure under fast penetration of electric vehicles. *World Electric Vehicle Journal, 11*(1), 18.
28. Nationalgrid. https://www.nationalgrid.com (Online; accessed 1 July 2020).
29. Soroudi, A., Amraee, T. (2013). Decision making under uncertainty in energy systems: State of the art. *Renewable and Sustainable Energy Reviews, 28*, 376–384.

Chapter 8
A Robust Optimization-Based Model for Smart Charging of PEV Under Multiple Uncertainties

Masoud Ahmadigorji, Majid Mehrasa, Seddik Bacha, and Ahmad Hably

8.1 Introduction

A. Status Quo, Challenges, and Outlook

The notion of plug-in electric vehicle (PEV) has universally attracted the attention of both academic and industrial individuals as a supportive choice to alleviate the techno-economic and especially environmental concerns [1, 2]. The clean energy and so non-polluting feature of the PEVs as well as the shortage of fossil fuel resources vindicate the integration of these emerging technologies with the operational planning of the electricity grids [3, 4]. Hence, it seems that the PEV penetration as a supple load/generation alternative is significantly increasing and so its future is encouraging [2, 5]. Nonetheless, the growing penetration of PEVs can inflict some operational difficulties based on local supervising and regulating policies of the grid operator [6]. In this context, despite the significant benefits of PEV operating in discharging mode at peak load times (which leads to peak load shaving, reliability enhancement, etc.), the non-controlled PEV charging trend in the light load period imposes a heavy load on the grid leading to voltage regulation, power quality, and accordingly security challenges [7]. However, both vehicle-to-grid (V2G) and grid-to-vehicle (G2V) characteristics of the PEVs are attractive issues that cannot be simply neglected in modern smart grids, wherein the mentioned capabilities are intelligently scheduled [8, 9]. On the other hand, incorporating both

M. Ahmadigorji (✉)
Department of Electrical Engineering, Nour Branch, Islamic Azad University, Nour, Iran

M. Mehrasa · S. Bacha
University of Grenoble Alpes, CNRS, Grenoble INP, G2ELAB, Grenoble, France

A. Hably
University of Grenoble Alpes, CNRS, Grenoble INP, GIPSA-Lab, Grenoble, France

V. Vahidinasab, B. Mohammadi-Ivatloo (eds.), *Electric Vehicle Integration via Smart Charging*, Green Energy and Technology,
https://doi.org/10.1007/978-3-031-05909-4_8

V2G and G2V modes of PEV operation into the planning/operation studies of the smart grids brings about additional complications and computational alternatives [10, 11].

Nowadays, the utilization of photovoltaic (PV) arrays as low-cost renewable energy resources has extensively absorbed the attention of researchers. The collaborative performance of these two technologies can remarkably enhance the overall operational purposes of energy management (EM) in smart microgrids [12–14]. In this way, an efficient EM tries to attain the most advantages of PV power in charging the electrical batteries of the PEV especially in light load periods. Conversely, the PEV operation in discharge mode within the time span of peak electricity price, where the load demand is ordinarily in peak condition, can enhance the technical/economic/environmental goals of the EM strategy in modern smart grids [15–19].

B. Literature Review

Many recently published research works address the EM approaches by various objective functions including cost minimization, benefit maximization, power balancing, peak load shaving, and so on [20–37]. Among the literature, some research works have focused on utilizing the PV potential aiming to attain both low-cost load supply and smart PEV charging [20–33]. On the other hand, various well-known optimization programming including both numerical optimization and evolutionary algorithms are widely used to achieve the most optimal EM mechanism [20–37]. Despite the valuable research work performed in the literature, two major issues should be taken into account within the EM-based PEV smart charging plan. First, the EM optimization procedure, as seen in the literature, has been performed from the system operator viewpoint while the other parties such as PEV users must be incorporated with the developed model. Secondly, it should be noted that the PEV smart charging is subject to the various uncertainty resources in practice. In this context, the deterministic smart charging framework in which the uncertainties are fully ignored cannot deal with the future probable deviation of the uncertain variables from their forecasted (expected) values. For this reason, an effective robust model withstanding the severe fluctuating of the uncertain parameters around their anticipated values (as an operational troublesome in real-time EM) seems to be so crucial.

C. Contribution

In this chapter, an efficient minimum-cost EM model is proposed to attain a smart PEV charging strategy coordinated with possible advantages of the PV arrays along a day. From viewpoint of the charging station owner, the revenue earned by injecting the PEV output in discharging mode (especially during peak load interval) is desirable to be maximized. On the other hand, the system operator tries to adequately supply the load demand by minimizing the cost of power purchased from the main grid. Without loss of generality, the system operator and the charging station owner are considered the same in this research. In this case, the common interest is considered as minimizing the cost of power procured from the main grid based on efficient exploiting of the PV daily generation together with smart PEV contribution in both charging/discharging modes.

Additionally, the essential role of PEV user is incorporated with the suggested model such that the charging/discharging pattern is optimally determined relying on the input data received from him/her before reaching the station. Hence, the proposed charging plan is on the basis of cooperative interaction between the station owner and PEV user considering the key role of PEV user (i.e., agreement between all parties). Within the proposed charging model, the PEV user can announce his/her arrival and expected departure time together with the initial state of charge (SOC) of the battery and the desired final one. Also, the battery characteristics as another essential data to apply the technical constraints are delivered from the PEV user. Based on hourly profiles of the load power, PV power, and also electricity price (available input data), the proposed EM model seeks to discover the minimum-cost PEV charging schedule during the time interval between its arrival and departure. Afterward, the optimal charging plan is announced to the PEV user before reaching the charging station. In practice, the entire cooperative infrastructure of data sending/receiving between the station owner and PEV user can be carried out by relevant applications installed on smart phones or other communication devices.

The PEV operation in the mode of charging/discharging is highly dependent on the effectual input parameters such as electricity price, extractable PV power (based on climate condition), and load power behavior. The mentioned parameters encounter inherent uncertainties in practice. Therefore, it is essentially required to use a proficient approach to cope with the uncertain behavior of such parameters. Based on this issue, the notion of information gap decision theory (IGDT) is incorporated with the proposed smart charging framework to investigate the challenging influence of uncertain variables on the decision-making process. The main capability of the formulated IGDT-based robust model is to immunize the PEV charging plan against any probable oscillation of the uncertain variables within their robust regions. Since the uncertain variables separately compete to maximize their own robust regions, the presented IGDT-based model is in the form of a multi-objective optimization problem. Accordingly, multi-objective particle swarm optimization (MOPSO) is utilized to give a set of non-dominated Pareto solutions. To reach the best compromise solution, a fuzzy satisfaction rule is introduced.

The chapter does not focus on a specific part, such as IGDT, ignoring the other parts. Emphasizing that the IGDT concept is not the only uncertainty modeling approach, the current chapter aims to highlight the specific capability of this technique to present a set of robust outcomes protected against the worst-case realization of the uncertain parameters throughout their robust regions. Hence, the IGDT is not the just contribution of this chapter and is only utilized as an efficient approach to handle the non-repeatable or low-frequency uncertainty resources considered in the proposed non-deterministic PEV charging model.

To further highlight the overall research work carried out in this chapter, see the graphical abstract depicted in Fig. 8.1.

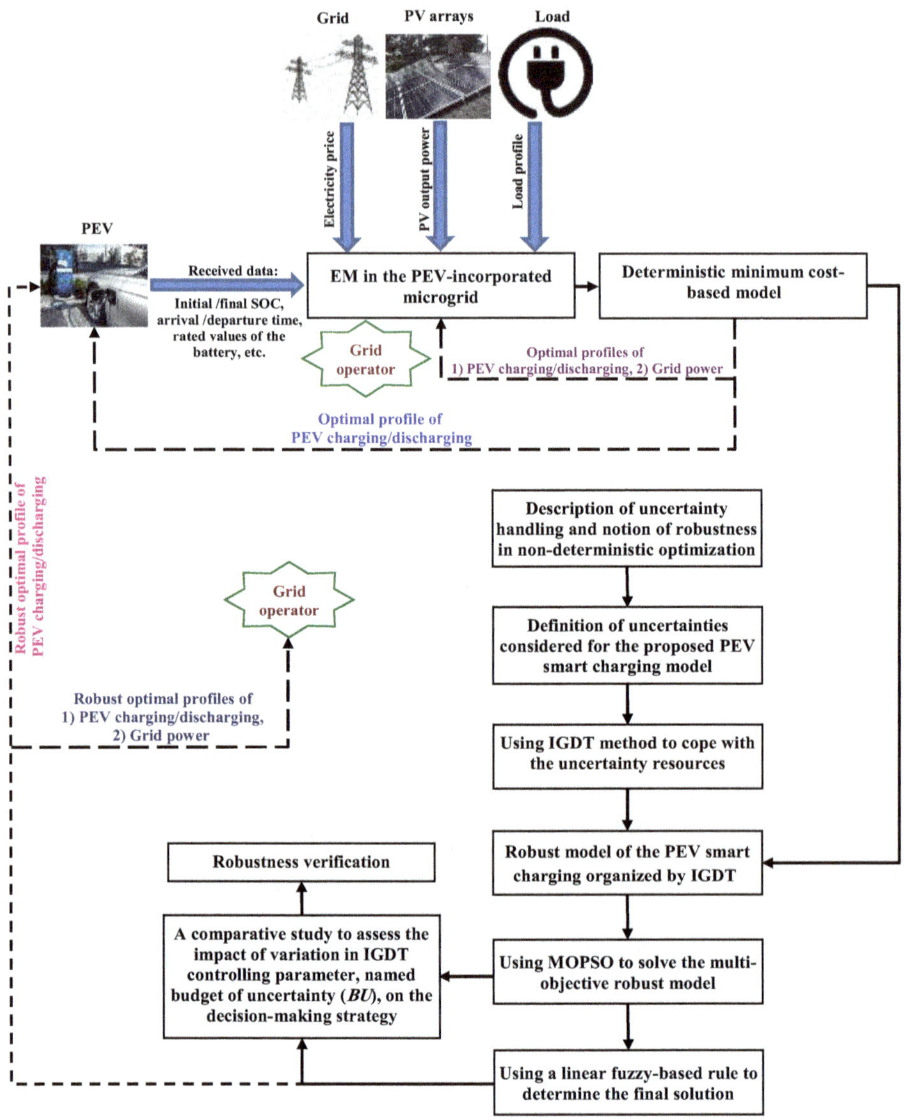

Fig. 8.1 The graphical abstract of the research work performed in this chapter

D. Chapter Organization

The rest of the chapter is organized as follows: In Sect. 8.2, the mathematical formulation of the deterministic model considered for the smart charging of PEV is presented. Then, the notion of IGDT, the proposed IGDT-based robust model for PEV smart charging, the MOPSO algorithm, and also the fuzzy satisfaction method are distinctly described in Sect. 8.3. The numerical results obtained by simultaneously applying the proposed IGDT-based charging model, the MOPSO algorithm as well as the fuzzy satisfaction method to a test microgrid are presented and discussed in Sect. 8.4. Finally, Sect. 8.5 encompasses the conclusion remarks.

8.2 Mathematical Representation of the Deterministic PEV Smart Charging

The proposed PEV charging strategy aims to minimize the daily total cost paid for purchase of the electricity from the main grid. Therefore, the mathematical formulation of the suggested model in the case of no uncertainty consideration (i.e., deterministic model) can be presented as follows:

$$OF_{det} \equiv \min cost_{grid}^{pur} \tag{8.1}$$

$$cost_{grid}^{pur} = \sum_{t=0}^{24} \left(GP_t \times EP_t\right) \tag{8.2}$$

where $\mathrm{OF}_{\mathrm{det}}$ and $\mathrm{cost}_{\mathrm{grid}}^{\mathrm{pur}}$ in (8.1) stand for the deterministic objective function and cost of power purchased from the main grid, respectively. As seen from (8.1), the main focus of the proposed PEV-incorporated EM model is to minimize the cost of power procured from the main grid. The $\mathrm{cost}_{\mathrm{grid}}^{\mathrm{pur}}$ value is calculated using (8.2) where GP_t and EP_t respectively denote the values of power supplied by the main grid and the electricity price in hour t.

8.2.1 Constraints

The optimization procedure of the proposed charging strategy is exposed to the following equality and inequality constraints:

- The hourly power balance should be always preserved:

$$GP_t = LP_t - PVP_t - \underbrace{(\mathrm{SOC}_t - \mathrm{SOC}_{t-\Delta t}) \times \left(En_{PEV}^{rated}/\Delta t\right)}_{EVP_t} \quad t = \{0, 1, \ldots, 24\} \tag{8.3}$$

where LP_t, PVP_t,and EVP_t represent the load power, PV power, and PEV charging/discharging power in hour t (i.e., end of period Δt), respectively; En_{PEV}^{rated} is the rated capacity of PEV. If the minimum time period of remaining in charging/discharging mode of PEV is considered to be 1 h (i.e., $\Delta t = 1$), the equality constraint of (8.3) can be rewritten as below:

$$\mathrm{GP}_t = \mathrm{LP}_t - \mathrm{PVP}_t - \underbrace{(\mathrm{SOC}_t - \mathrm{SOC}_{t-1}) \times \left(\mathrm{En}_{\mathrm{PEV}}^{\mathrm{rated}}\right)}_{\mathrm{EVP}_t} \qquad t = \{0, 1, \ldots, 24\} \tag{8.4}$$

According to (8.4), by comparing SOC_t and SOC_{t-1}, three operation status for a PEV can be enumerated, as follows:

$$\begin{cases} \mathrm{SOC}_t > \mathrm{SOC}_{t-1} & ; \quad \text{charging mode} \\ \mathrm{SOC}_t < \mathrm{SOC}_{t-1} & ; \text{discharging mode} \\ \mathrm{SOC}_t = \mathrm{SOC}_{t-1} & ; \quad \text{idle model} \end{cases} \tag{8.5}$$

The negative sign of EVP_t in (8.4) is considered in accordance with Kirchhoff current law (KCL). Based on KCL, all input currents to an electrical node are positive-signed and the output ones have negative sign. Thus, as illustrated in Fig. 8.2, if the system operator is structurally considered to play the role of a centralized electrical node, the entire input powers are considered to have positive sign and contrarily the output powers are negative-signed. By this rule, if $\mathrm{SOC}_t > \mathrm{SOC}_{t-1}$, it means that the PEV is in the charge mode and so acts as a load receiving power from the main grid. Thus, the EVP_t sign should be originally negative (from the system operator viewpoint) in charging mode based on KCL. Hence, by multiplying the negative sign, inscribed behind the EVP_t, into a negative value of EVP_t in charging mode, a positive value for the EVP_t would be obtained reasonably leading to increment in the grid power. The same justification can be declared for the PEV discharging condition.

It is worth mentioning that without loss of the generality, the charge/discharge efficiency of PEV is considered to be 100% (i.e., ideal battery), which does not restrict the performance of the proposed model.

- The battery SOC during the time interval between the PEV arrival and departure varies according to the following pattern:

$$\mathrm{SOC}_{t \le t_{arr}} = \mathrm{SOC}_{arr} \tag{8.6}$$

$$\mathrm{SOC}_{t \ge t_{dep}} = \mathrm{SOC}_{dep} \tag{8.7}$$

$$\mathrm{SOC}_{t_{arr} < t < t_{dep}} = \mathrm{SOC}_{t-1} + \left(EVP_t / En_{PEV}^{rated}\right) \times (\Delta t = 1) \tag{8.8}$$

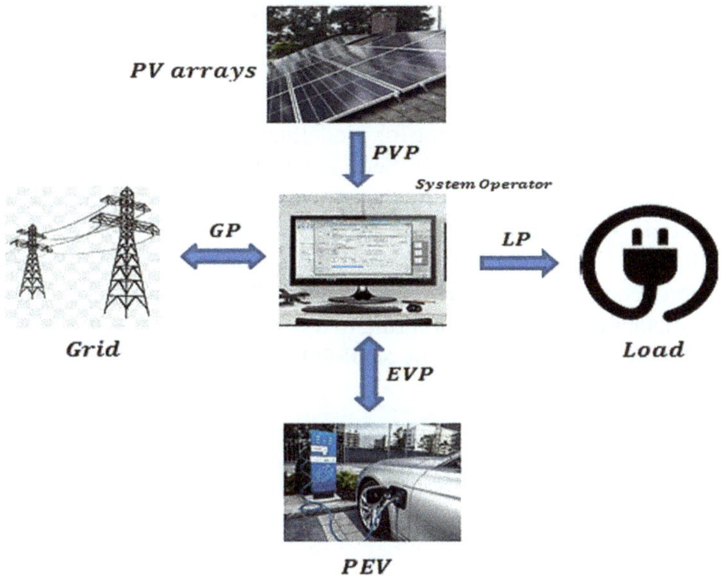

Fig. 8.2 The schematic of a hypothtical PEV-incorporated microgrid

where SOC_{arr} and SOC_{dep} are the arrival and desired departure SOC, respectively, collected from the PEV user before his/her arrival to the charging station. Also, t_{arr} and t_{dep} are the arrival/departure time of PEV to/from the charging station.

- The PEV power, the battery SOC, and also the stockpiled energy at the end of hour t should be permanently within their corresponding permissible range (declared by the manufacturer) aiming to prohibit the loss of battery life:

$$P_{PEV}^{min} \leq EVP_t \leq P_{PEV}^{max} \tag{8.9}$$

$$\mathrm{SOC}_{min} \leq \mathrm{SOC}_t \leq \mathrm{SOC}_{max} \tag{8.10}$$

$$\varphi_{PEV}^{min} \leq (EVP_t \times (\Delta t = 1)) \leq \varphi_{PEV}^{max} \tag{8.11}$$

$$\varphi_{PEV}^{min} = \mathrm{SOC}_{min} \times \mathrm{En}_{PEV}^{rated}; \varphi_{PEV}^{max} = \mathrm{SOC}_{max} \times \mathrm{En}_{PEV}^{rated} \tag{8.12}$$

where SOC_{min} and SOC_{max} are the minimum and maximum permissible SOC of the PEV battery. Also, φ_{PEV}^{min} and φ_{PEV}^{max} are respectively the minimum and maximum stockpiled energy of PEV battery based on SOC_{min} and SOC_{max}, respectively.

8.3 The Proposed IGDT-Based Model for Robust Smart PEV Charging

In the following, the IGDT concept for modeling the uncertainty resources of the proposed model is first described in Sect. 8.3.1. Then, the proposed IGDT-based model for PEV smart charging is mathematically described in Sect. 8.3.2. As the formulated IGDT-based problem is in the form of a multi-objective optimization problem, the step-by-step performance of the multi-objective PSO (MOPSO) as the solution approach to the mentioned problem is explicated in Sect. 8.3.3. Ultimately, the fuzzy satisfaction-based decision rule to determine the final solution among the Pareto optimal solutions generated by MOPSO is explained in Sect. 8.3.4.

8.3.1 The Information Gap Decision Theory (IGDT)

There are several techno/economic types of uncertainty resources in both planning and operation studies of the power systems required to be appropriately modeled. To this end, the well-known probability distribution function (PDF) [38–40] for probabilistic methods and fuzzy membership function (FMF) [41–43] for possibilistic frameworks have been frequently utilized in recent research works in the area. However, in some cases, the PDF of such variables is not available or not sufficient for accurate uncertainty modeling. Additionally, the variation trend of some uncertain variables does not follow a well-known PDF. On the other hand, an FMF needs the previous experiences of expert systems which are not simply accessible in some conditions. Therefore, a proficient uncertainty handling method such as IGDT can be a beneficial tool aiming to deal with such uncertainty resources.

Within the IGDT performance, the main target is to maximize the robustness of the decision-making outcomes. In this way, the permissible deviation of the uncertain variables from their forecasted (most likely) values is limited to a boundary value known as robust region. Based on this issue, the possible realization of the uncertain variables is within their own robust region. In other words, the IGDT tends to enhance the resistance of the outputs of a non-deterministic problem against any possible realization of the uncertain variables all over their own robust regions [44–46]. In this context, the IGDT tries to concurrently maximize the boundary values of the robust regions pertaining to all uncertain variables leading to achieve the most robust solution. The robust solution of the IGDT is immunized against any value of the uncertain variables realized within their own robust region.

The concept of robustness in scheduling/operation problems directly depends on the techno-economic policies, expectation, priorities as well as the conservatism level of the decision-maker. The most common robustness criterion broadly utilized in non-deterministic optimization problems is not exceeding the non-deterministic objective function from a predetermined value [44–49]. The mentioned value is often a multiple of the deterministic objective function obtained by neglecting the uncertainty resources (i.e., when all variables/parameters are certain). On the other hand, the declared multiple is a function of a controlling user-defined parameter known as budget of uncertainty (*BU*). The *BU* selection bounding the robust cost in IGDT is highly reliant on how much deviation of the non-deterministic cost from the deterministic one (i.e., robustness) as well as the prediction error occurred for the decision outcomes is practically tolerable by the decision-maker. The more discrepancy between the robust and deterministic costs permitted by the higher BU value, the more robustness as well as the more extended robust regions for the uncertain variables is reached. Furthermore, the higher *BU* value leads to the problem risk augmentation and so more deviation of the uncertain variables from their own predicted values throughout their pertinent robust regions. Nevertheless, the decision-maker should be careful about the appropriate definition of the *BU* value in practical uncertain environments to prevent the overestimation of the decision variables. Thus, it can be stated that the main mission of IGDT is to maximize the robust regions of the uncertain variables under the two following restrictive criteria: (1) the entire realization of the uncertain variables must be within their robust region, and (2) the robustness of the problem, depending on the *BU* parameter, should be always preserved.

There are numerous types of uncertainty modeling in IGDT. The widely used envelope bound method is utilized in this chapter [45–49]. To demonstrate the mathematical formulation of the envelope model, suppose an optimization problem with the set of decision variables ***DV*** and uncertain variables ***UV***. If the set of boundary (envelope) values pertaining to the robust regions of the ***UV*** is demonstrated by $\boldsymbol{\beta}$, the envelope bound method identifying the robust region of ***UV*** (i.e., ***rr***) can be mathematically presented as below:

$$UV \in \mathrm{rr}\left(\beta, \widetilde{UV}\right) \tag{8.13}$$

$$\mathrm{rr}\left(\beta, \widetilde{UV}\right) = \left|\frac{UV - \widetilde{UV}}{\widetilde{UV}}\right| \leq \beta \tag{8.14}$$

where $\widetilde{UV}$ is the anticipated (most probable) value of ***UV***. From (8.14), the uncertain variables pertaining to ***UV*** can take different values within their own robust region bounded by the related $\boldsymbol{\beta}$ value. Thus, the variation range of ***UV*** considering the restricting value of $\boldsymbol{\beta}$ can be illustrated as follows:

$$(1 - \boldsymbol{\beta}) . \widetilde{\boldsymbol{UV}} \leq \boldsymbol{UV} \leq (1 + \boldsymbol{\beta}) . \widetilde{\boldsymbol{UV}} \tag{8.15}$$

The main task of IGDT is to maximize the set of $\boldsymbol{\beta}$ leading to attain a set of decision variables hedged against all possible realization of the uncertain variables throughout their own robust regions. The more the $\boldsymbol{\beta}$ value, the more robustness of the decision-making results (i.e., $\boldsymbol{DV}$).

As previously explained in this section, the robustness considered in this chapter is not exceeding the non-deterministic (robust) objective function from a multiple of the deterministic one. For the sake of better insight, let consider $f_{\text{rob}}(\boldsymbol{UV}, \boldsymbol{DV})$ as the non-deterministic objective function value. It is evident that increasing the value of the uncertain variables pertaining to the set of $\boldsymbol{UV}$ brings to worsening the $f_{\text{rob}}(\boldsymbol{UV}, \boldsymbol{DV})$. Thus, the worst-case value for $f_{\text{rob}}(\boldsymbol{UV}, \boldsymbol{DV})$ to ensure the highest robustness of the problem occurs when the entire uncertain variables of $\boldsymbol{UV}$ stick to their upper bounds (i.e., $\boldsymbol{UV} = (1 + \boldsymbol{\beta}) .\widetilde{\boldsymbol{UV}}$). Hence, the overall IGDT mathematical formulation can be depicted by (8.16)–(8.18):

$$\max_{\boldsymbol{DV}} \boldsymbol{\beta} \left(\boldsymbol{DV}, f_{det} \left(\boldsymbol{DV}\right)\right) \tag{8.16}$$

$$\boldsymbol{UV} \in \boldsymbol{rr}\left(\boldsymbol{\beta}, \widetilde{\boldsymbol{UV}}\right) \tag{8.17}$$

$$f_{rob} \left(\boldsymbol{DV}, \left(\boldsymbol{UV} = (1 + \boldsymbol{\beta}) .\widetilde{\boldsymbol{UV}}\right)\right) \leq (1 + BU) . f_{det} \left(\boldsymbol{DV}\right) \tag{8.18}$$

where BU, generally known as the budget of uncertainty, is a confining parameter quantified according to the preference and also priorities of the decision-maker. In other words, the BU value (usually known as IGDT freedom degree) controls the IGDT performance to optimally specify a set of $\boldsymbol{DV}$ such that the possible biggest size of $\boldsymbol{rr}$ satisfying the robustness is achieved.

8.3.2 The Proposed IGDT-Based PEV Smart Charging

The main idea behind the IGDT application as the uncertainty handling approach of this chapter is its specific capability to cope with the low-frequency (or non-repeatable) uncertainty resources. In other words, the IGDT is a useful tool to deal with the uncertainty resources having low historical data or no definite predictable behavior. For such uncertainty resources, the probability distribution function (PDF) or fuzzy membership function cannot be simply organized. The latter customarily relies on the knowledge of the expert operators who have adequate information about the previous realizations of the uncertain variables. However, due to unexpected events, these operators are not always accessible in practice or their information cannot be fully guaranteed. The PEV smart charging model, presented in this chapter, is exposed to three uncertainty recourses including load power, PV output power, and the electricity price. Moreover, the proposed robust model tries to attain the optimal hourly profile of the PEV power based on the hourly uncertain input data. In this case, only 24 available samples for each of uncertain data are not so sufficient and helpful to constitute an accurate PDF. Furthermore, the electricity price does not intrinsically follow a definite PDF or even fuzzy membership function due to its occasional spikes occurring based on political, social, or cultural events, as previously seen in the universal power markets. In this circumstance, the accurate modeling of the mentioned uncertainty resources by means of the probabilistic approaches (i.e., Monte Carlo simulation, point estimation method, etc.) or fuzzy-based possibilistic methods is a non-trivial task (if not impossible). Thus, the notion of IGDT presented in Sect. 8.3.1 is applied to the deterministic model of (8.1)–(8.12) aiming to find a robust set of decision outcomes. To develop the IGDT-based PEV charging model, first consider the set of uncertain variables, set of robust regions, and set of bounding (envelope) values respectively depicted in (8.19)–(8.21):

$$\boldsymbol{UV} = \{\boldsymbol{LP_t}, \boldsymbol{PVP_t}, \boldsymbol{EP_t}\} \tag{8.19}$$

$$\boldsymbol{rr} = \left\{\boldsymbol{rr_{lp}}, \boldsymbol{rr_{pv}}, \boldsymbol{rr_{pr}}\right\} \tag{8.20}$$

$$\boldsymbol{\beta} = \left\{\boldsymbol{\beta_{lp}}, \boldsymbol{\beta_{pv}}, \boldsymbol{\beta_{pr}}\right\} \tag{8.21}$$

In practice, there are numerous uncertainty resources to be integrated with the proposed deterministic model of (8.1)–(8.12). Among them, the three prevalent load power, PV generation, and electricity price are considered as the main uncertainty resources, which are mathematically modeled by means of the IGDT approach. Without loss of the generality, it is assumed that the time of PEV arrival/departure to/from the charging station is certain and so in accordance with the desire of PEV user. Based on the envelope bound method, the uncertain variables can take different values throughout their own robust region. Thus, the robust region of load power, PV power, and electricity price (i.e., ***UV***) can be demonstrated with (8.22)–(8.24), respectively:

$$\boldsymbol{rr_{lp}} = \left| \frac{LP_t - \widetilde{LP}_t}{\widetilde{LP}_t} \right| \leq \boldsymbol{\beta_{lp}} \tag{8.22}$$

$$\boldsymbol{rr_{pv}} = \left| \frac{PVP_t - \widetilde{PVP}_t}{\widetilde{PVP}_t} \right| \leq \boldsymbol{\beta_{pv}} \tag{8.23}$$

$$\boldsymbol{rr_{pr}} = \left| \frac{EP_t - \widetilde{EP}_t}{\widetilde{EP}_t} \right| \leq \boldsymbol{\beta_{pr}} \tag{8.24}$$

Regarding the IGDT definition, the maximum value of the objective function associated with the non-deterministic model occurs while all the uncertain variables experience the worst-case value within their robust regions. In this context, the worst-case realized values for $\boldsymbol{UV}$ members based on (8.22)–(8.24) are $\left(1+\beta_{lp}\right).\widetilde{LP}_t$ for load power, $\left(1-\beta_{pv}\right).\widetilde{PVP}_t$ for PV power, and $\left(1+\beta_{pr}\right).\widetilde{EP}_t$ for electricity price. Thus, introducing $\boldsymbol{DV}=\{\boldsymbol{EVP_t},\boldsymbol{GP_t}\}$, the proposed IGDT-based PEV charging model can be formulated as below:

$$\max\ \left(\boldsymbol{\beta}_{lp},\boldsymbol{\beta}_{pv},\boldsymbol{\beta}_{pr}\right) \tag{8.25}$$

Subject to:

$$OF_{rob}\equiv \mathrm{cost}_{grid}^{pur}\ (\boldsymbol{GP},\boldsymbol{SOC},\boldsymbol{LP},\boldsymbol{PVP},\boldsymbol{EP})\leq(1+BU).OF_{det} \tag{8.26}$$

$$\mathrm{cost}_{grid}^{pur}=\sum\nolimits_{t=0}^{24}\ \left(GP_t.\left(1+\beta_{pr}\right).EP_t\right) \tag{8.27}$$

$$GP_t=\left(1+\beta_{lp}\right).LP_t-\left(1-\beta_{pv}\right).PVP_t-EVP_t \tag{8.28}$$

$$(8.6)\text{–}(8.12) \tag{8.29}$$

Since reduction in PV power due to climate change brings to increase in OF_{rob}, the worst-case realized value of PVP_t becomes$\left(1-\beta_{pv}\right).\widetilde{PVP}_t$. The proposed IGDT-based charging model of (8.25)–(8.29) is inherently a multi-objective optimization problem. For this reason, the multi-objective PSO (MOPSO) algorithm is utilized to solve the formulated model. Since the MOPSO generates a set of non-dominated solutions on the Pareto front, a fuzzy satisfaction decision-making technique is employed to select the final solution among the Pareto solutions. The performance process of the MOPSO and fuzzy satisfaction method are described in the following.

8.3.3 Multi-objective Particle Swarm Optimization (MOPSO)

In the following, the overall performance of the particle optimization (PSO) is concisely reviewed in Sect. 8.3.3.1. Afterward, the multi-objective form of PSO namely MOPSO is described in Sect. 8.3.3.2.

8.3.3.1 Concise Review of PSO Algorithm

Particle swarm optimization (PSO) is a population-based optimization algorithm motivated by the collective movement of birds or fish platoon. In this algorithm, the set of candidate solutions called particles (which simulate the birds or fish behavior) move within a feasible search space of the problem aiming to reach the optimal or near to optimal solution. The initial classic PSO has been introduced by Kennedy and Eberhart [50] and then, its enhanced variants have frequently been utilized in recent researches of the power system [51–55]. Mathematically speaking, each particle of PSO plays the role of a candidate solution for the optimization problem. The position and velocity associated with ith particle of PSO in a m-dimensional optimization problem can be indicated by $PX_i = \left[px_i^1, px_i^2, \ldots, px_i^m\right]$ and $V_i = \left[v_i^1, v_i^2, \ldots, v_i^m\right]$, respectively.

Within the PSO evolution procedure, each particle moves toward a path generated by the combinatorial interaction of three experiences recorded in the previous iteration. These experiences comprise the previous position of particle, the best experience recorded by that particle (*PB*), and the best position experienced among the population (*GB*). Accordingly, the iterative movement (evolution) of the PSO particles can be formulated as follows:

$$V_i^{it+1} = \alpha.V_i^{it} + C_1.rd_1.\left(PB_i^{it} - PX_i^{it}\right) + C_2.rd_2.\left(GB^{it} - PX_i^{it}\right) \qquad i \in \{1, .., N_{par}\} \tag{8.30}$$

$$PX_i^{it+1} = PX_i^{it} + V_i^{it+1} \tag{8.31}$$

where N_{par} is the number of particles (population). Additionally, C_1, C_2, and α are three weighting coefficients designated respectively for controlling the particle movement toward the GB, the particle's previous PB, and previous velocity. With the aim of more search efficiency, the factor α begins with the higher value in initial iterations and consequently decreases to the lower value during the PSO evolution procedure. Note that typically $C_1 + C_2 = 4$; rd_1 and rd_2 are two random variables between 0 and 1 with uniform distribution.

The update process of (8.30) and (8.31) is continued for all population members until the stopping criterion (generally maximum number of iterations ($it = it_{\max}$)) is reached. The *GB* of the last iteration is selected as the best solution of the PSO.

8.3.3.2 The Concept of Dominance in a Multi-objective Problem

Suppose we have a multi-objective problem with n decision variables and *nobj* objectives. Without loss of the generality, assume a maximization-based multi-objective optimization problem. The mathematical representation of such problem can be presented as follows:

$$F(X) = \left[f_1(X), f_2(X), \ldots, f_{nobj}(X)\right] \tag{8.32}$$

$$X = [x_1, x_2, \ldots, x_n] \tag{8.33}$$

$$\left\{\text{EQ}(X) = \overline{0}, \quad \text{IEQ}(X) \leq \overline{0}\right\} \tag{8.34}$$

where EQ(X) and IEQ(X) stand for the set of equality and inequality constraints. In such problems, solution X_1 dominates the solution X_2 if the two following conditions are simultaneously satisfied:

$$\forall b \in \{1, \ldots, nobj\} \Rightarrow \quad f_b(X_1) \geq f_b(X_2) \tag{8.35}$$

$$\exists b^* \in \{1, \ldots, nobj\} \Rightarrow f_{b^*}(X_1) > f_{b^*}(X_2) \tag{8.36}$$

8.3.3.3 The MOPSO Step-by-Step Implementation

The GB in single-objective PSO is substituted by a set of non-dominated solutions within the evolution procedure of multi-objective PSO (MOPSO). Based on this issue, the MOPSO has an external archive called "repository" which is a space for saving the non-dominated solutions (i.e., Pareto solutions) in every iteration. Consequently, the performance steps of the MOPSO to generate a set of non-dominated solutions are presented as follows [56–58]:

Step 1- Set the iteration number to zero (i.e., $it = 0$). Then, randomly generate an initial population regarding the permissible range of decision variables as well as constraints' satisfaction.

Step 2- Calculate the value of all objective functions associated with every particle of population. In this step, the *PB* for each particle is as initially created in step 1.

Step 3- Select a predefined number of particles as non-dominated solutions and so store them in the repository. This issue is carried out based on the dominance concept explained in Sect. 8.3.3.2.

Step 4- For each remaining particle, one of the repository members is selected as the "leader". To do so, the Pareto front is divided into some adjacent cells (i.e., matrix configuration) by means of grid formation proposed in [56]. By doing so, each cell may include one, more than one or even no Pareto optimal solution. In this way, the leader selection probability of cell i having n_i member $\left(P_i^{\text{ls}}\right)$ can be computed based on the following Boltzmann function:

$$P_i^{\mathrm{ls}} = \frac{\exp\left(-\sigma.n_i\right)}{\sum_{n_j} \exp\left(-\sigma.n_j\right)} \tag{8.37}$$

where n_i and σ (a value in the range of (0,1)) are the number of repository members located in cell i and leader selection pressure, respectively. Now, by having all cells' leader selection probability, one cell is randomly specified for each particle based on the Roulette Wheel selection method [59]. It is evident that the less the cell member, the more probability of its leader selection probability (due to diversity preservation). One member of the selected cell is randomly chosen as the leader.

Step 5- The iteration number is increased by one (i.e., $it = it + 1$). The velocity and position of each particle are updated identically to the approach explained in (8.30) and (8.31) while *GB* in (8.30) is replaced by the leader. If the current position of the particle dominates its previous *PB*, the *PB* of that particle is replaced by the current position. If any of the current and previous positions of the particle cannot dominate each other, the new *PB* is randomly selected between the two mentioned positions.

Step 6- Check the dominance condition for new position of the particles and then, add the new non-dominated particles to the repository of the previous iteration. Once more, check the new repository for the dominance probability of some members. By doing so, the dominated members of the current repository should be discarded.

Step 7- If the repository size exceeds the predefined value, the extra members should be removed. To do this, contrary to the approach presented in step 4, the more a cell members, the more the removing probability of members pertaining to that cell. Accordingly, the Boltzmann function of (8.37) is converted to (8.38) in this case:

$$P_i^{\mathrm{rem}} = \frac{\exp\left(-\rho.n_i\right)}{\sum_{n_j} \exp\left(-\rho.n_j\right)} \tag{8.38}$$

where P_i^{rem} and ρ (a value in the range of (0,1)) is the removing probability of cell i and removing selection pressure, respectively. By possessing the entire cells' removing probability, one cell is randomly specified based on the Roulette Wheel selection. Finally, one member of the selected cell is randomly removed. This process is continued until the predefined size of the repository is realized.

Step 8- Check the stopping condition. If the maximum iteration of MOPSO is reached, the final updated repository is determined as the Pareto optimal solutions. Otherwise, the evolution procedure is continued by going back to step 4.

The overall performance of the MOPSO is schematically clarified in Fig. 8.3.

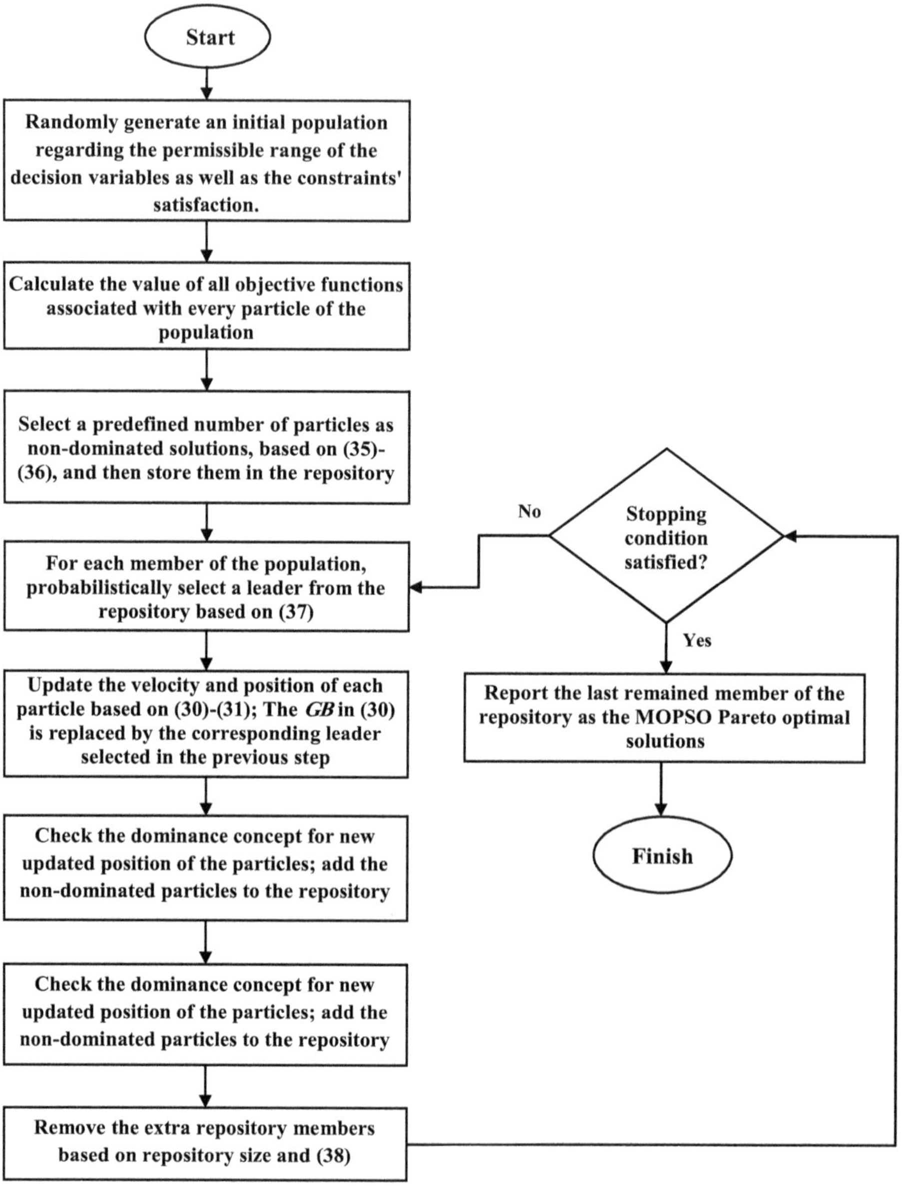

Fig. 8.3 The schematic illustration of the MOPSO performance

8.3.4 *Fuzzy Satisfaction Method*

The Pareto optimal solutions generated by a multi-objective solution approach can be potentially the final solution. Generally speaking, depending on predilection and priorities of the decision-maker, one of the Pareto optimal solutions is selected as the final outcome. However, while there is no priority between the objectives, the fuzzy satisfaction method can be a useful tool to choose the final solution [60].

Based on (8.32)–(8.34), assume b and np to be respectively the counter of each objective and each Pareto solution (i.e., $b = \{1, 2, \ldots, nobj\}$, $np = \{1, 2, \ldots, (N_{ps} =$ number of Pareto solutions$)\}$), respectively. Hence, the linear fuzzy memberships function ($\mu_{X_{np}}^{f_b}$) is calculated for solution X_{np} as follows:

$$\mu_{X_{np}}^{f_b} = \begin{cases} 1 & f_b\left(X_{np}\right) > f_{b,max} \\ \frac{f_{b,min} - f_b\left(X_{np}\right)}{f_{b,min} - f_{b,max}} & f_{b,min} \le f_b\left(X_{np}\right) \le f_{b,max} \\ 0 & f_b\left(X_{np}\right) < f_{b,min} \end{cases} \tag{8.39}$$

where $f_{b,\,min}$ and $f_{b,\,max}$ are the minimum and maximum values of f_b among the set of Pareto optimal solutions. For every X_{np}, the minimum satisfaction $\mu_{X_{np}}^{f_b}$ related to each objective function is determined. The X_{np} with maximum value of $\mu_{X_{np}}^{f_b}$ is selected as the best solution, as below:

$$\max_{np=1,\ldots,N_{ps}} \min_{b=1,\ldots,nobj} \mu_{X_{np}}^{f_b} \tag{8.40}$$

8.4 Numerical Results

8.4.1 *Input Data*

To assess the effectiveness of the proposed IGDT-based PEV smart charging, a set of load, PV arrays, and a PEV (as a flexible load/generation) is constituted in the form of a microgrid structurally similar to the Fig. 8.2. The daily profiles of load power, PV power as well as the electricity price are depicted in Figs. 8.4a, b and 8.5, respectively [61]. The technical specifications of the PEV (2011 Nissan Leaf) are presented in Table 8.1. Also, the setting parameters of MOPSO are shown in Table 8.2. The arrival and departure of SOC is assumed to be 0.4 and 0.8, respectively. For the sake of more clarification and better evaluation of the PEV influence on the daily cost, its arrival and departure times are assumed to be 4.00 and 23.00, respectively. In this way, the PEV is approximately accessible for the entire day and behaves as a motionless source of energy storage. It is worthwhile to mention that all the simulations are carried out in the programming environment of MATLAB 7.5 using the hardware set of a personal computer with a Pentium IV processor, clocking at 2.3 GHz, and 4 GB RAM.

8.4.2 The SOC and Power Analysis

In this section, the simulation results attained by applying the proposed IGDT-based charging model, MOPSO, and fuzzy satisfaction method for distinct *BU* values of {1, 1.25, 1.5, 1.75, and 2} are presented and discussed. By applying the proposed IGDT-based charging model along with MOPSO, 20 Pareto optimal solutions are generated for each *BU* having their own *rr* (i.e., rr_{lp}, rr_{pv}, and rr_{pr}). To select the final solution among the 20 obtained Pareto solutions, the fuzzy satisfaction method introduced in Sect. 8.3.4 is employed. The average simulation time to find the final compromise solution for each BU value is 30 seconds based on the utilized hardware set. For each non-zero *BU* value, the envelope bounds of rr_{lp}, rr_{pv}, and rr_{pr} (i.e., β_{lp}, β_{pv}, and β_{pr}) pertaining to each 20 optimal Pareto solutions are graphically demonstrated in Figs. 8.6, 8.7, and 8.8, respectively.

For the final solution determined by the fuzzy satisfaction method, the SOC variations together with the power profiles of the load, PV, grid, and PEV are presented in Figs. 8.9, 8.10, 8.11, 8.12, and 8.13 respectively for *BU* values of 1, 1.25, 1.5, 1.75, and 2. To figure out the daily trend of the SOC and PEV power, the maximum and minimum values of the electricity price must be taken into account as illustrated in Fig. 8.5. As shown in Figs. 8.9, 8.10, 8.11, 8.12, and 8.13, at the peak electricity prices (i.e., $t = 5, 20$), the PEV reaches the maximum discharging power injected into the grid for all *BU*s and so the SOC tends to the reduction prior to the peak times. In contrast, during the minimum price interval [7 h, 16 h], the proposed smart strategy enables the PEV to be at a charging state for all *BU* values. In addition, the maximum exploitation of the PV unit must be extremely paid attention in the case of different *BU* values. According to Figs. 8.9, 8.10, 8.11, 8.12, and 8.13, at $t = 12$ in which the PV unit generates its maximum power and also the electricity price is the least, the PEV operates at its maximum charging power for $BU = 1$, 1.25, 1.5, and 2. For these *BU* values, the grid supplies the extra power to the extent that the robust operation is established. However, at the same time for $BU = 1.75$, both load and grid completely consume the PV power (aiming to reduce the purchase cost) and the PEV power reaches zero (i.e., the PEV is in idle mode). As a comparison between the PEV power achieved from simulations for various *BU* values, the PEV power associated with the $BU = 1$ has more stable charging mode throughout the minimum price interval [7 h, 16 h] while the most fluctuations for charging strategy (i.e., PEV power) are obtained for $BU = 1.25$, 1.75, and 2, as observed from Figs. 8.9, 8.10, 8.12, and 8.13, respectively. Furthermore, the PEV power with $BU = 1.5$ is remained at charging mode less than other ones. It can be comprehended from Figs. 8.9, 8.10, 8.11, 8.12, and 8.13 that the proposed robust

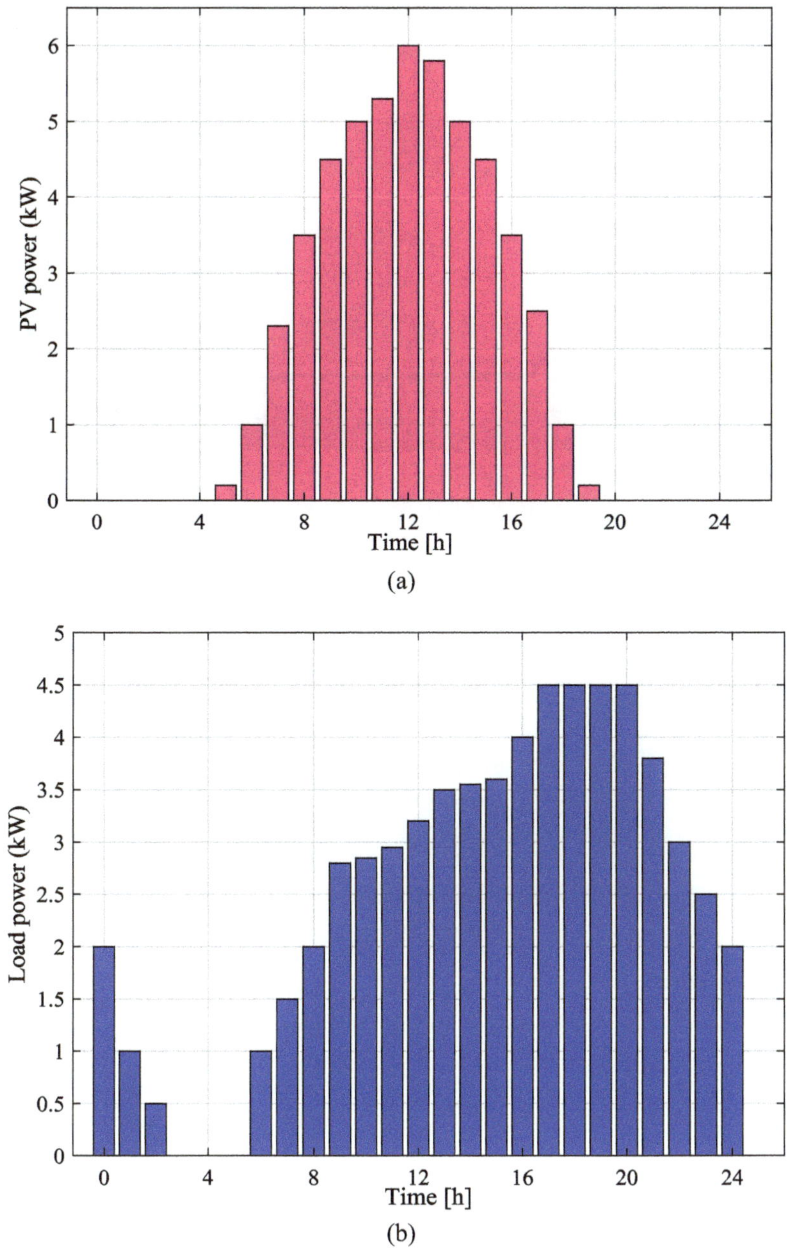

Fig. 8.4 Daily profiles of (**a**) PV power and (**b**) load power

charging strategy enables the PEV to have its specified SOC at arrival and departure times which are previously set to 40% and 80%, respectively.

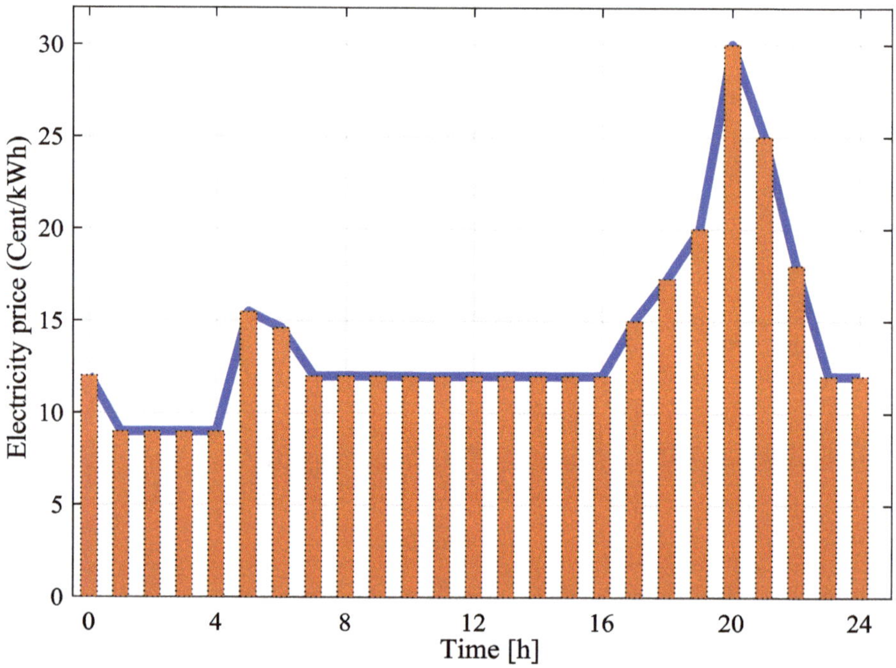

Fig. 8.5 Electricity price

Table 8.1 The technical specification of PEV

Symbol	Value	Unit
PVP_{nom} (rated capacity of PV arrays)	7	kW
$P_{PEV}^{min}/P_{PEV}^{max}$	−3.5/+3.5	kW
En_{PEV}^{rated}	24	kWh
SOC_{arr}/SOC_{dep}	0.4/0.8	–
SOC_{min}/SOC_{max}	0.2/0.9	–

Table 8.2 The setting parameters of MOPSO

Parameter	Value
Population size	100
Repository size	20
Maximum iteration ($iter_{max}$)	100
Leader selection pressure (σ)	0.4
Removing selection pressure (ρ)	0.4
C_1,C_2	2

8.4.3 Robustness Assessment

The robustness of the proposed smart charging model is interpreted based on Table 8.3 and Fig. 8.14. By increasing the *BU* value, the MOPSO highly attempts to extend the envelope bounds of the robust regions as much as possible. Generally

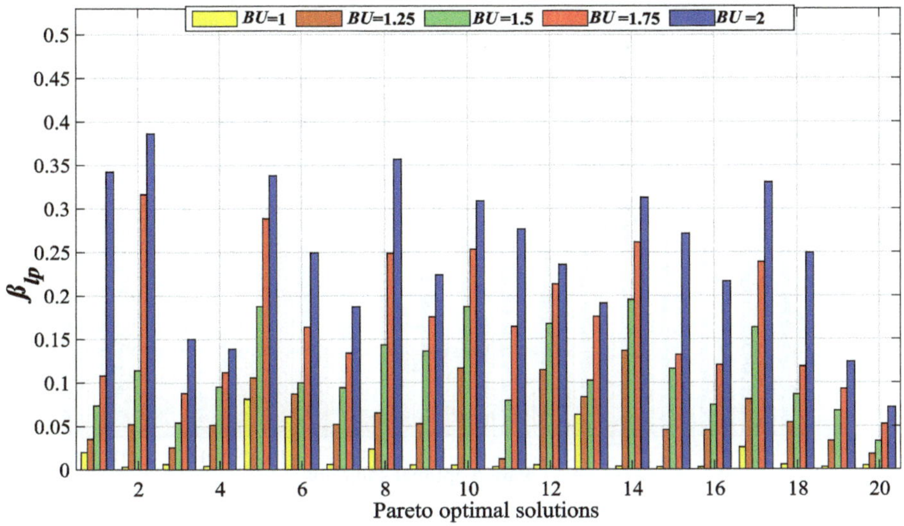

Fig. 8.6 The β_{lp} values of the 20 Pareto optimal solutions obtained for $BU = 1$, 1.25, 1.5, 1.75, and 2

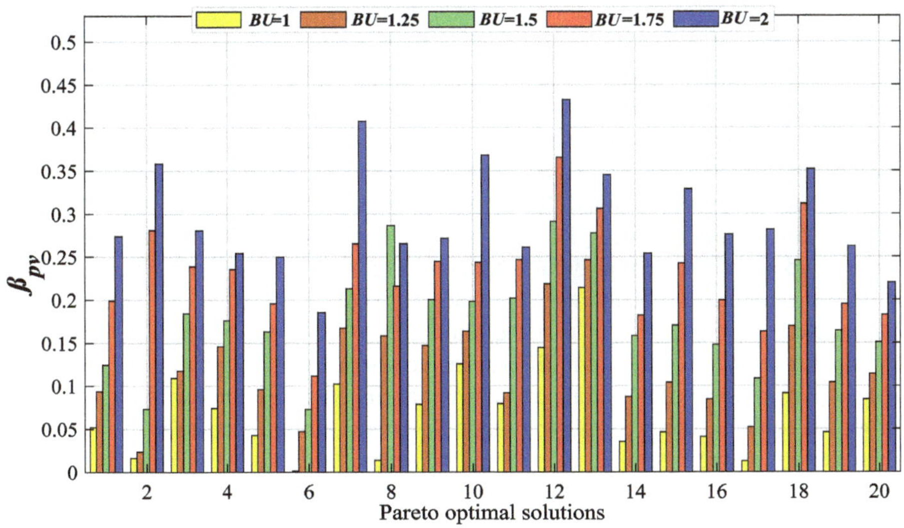

Fig. 8.7 The β_{pv} values of the 20 Pareto optimal solutions obtained for $BU = 1$, 1.25, 1.5, 1.75, and 2

speaking, it is expected that increasing of the *BU* value allows increase in OF_{rob} value by simultaneous augmentation of all envelope bounds. In this context, Table 8.3 shows that the higher envelop bounds of β_{lp}, β_{pv}, and β_{pr} are attained while the *BU* increases leading to more robustness of the decision variables against the fluctuation of the uncertain variables. Moreover, it is evident from Table 8.3 that

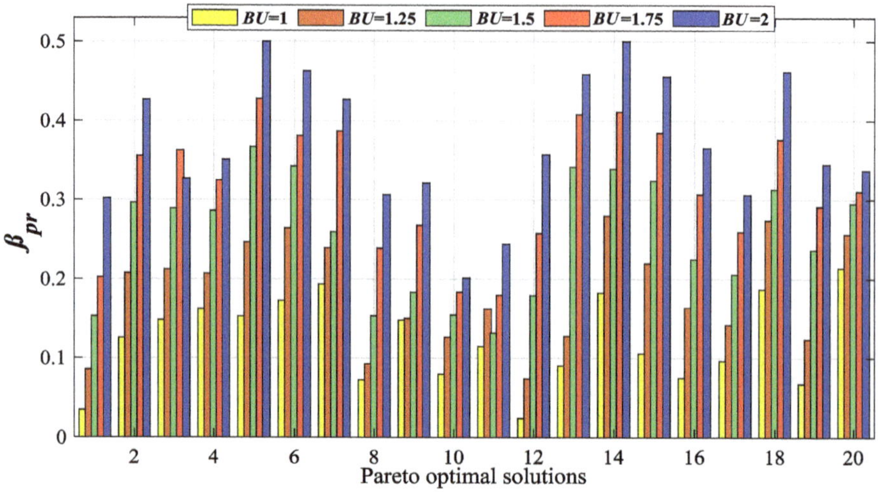

Fig. 8.8 The β_{pr} values of the 20 Pareto optimal solutions obtained for $BU = 1, 1.25, 1.5, 1.75$, and 2

the higher values of the envelope bounds result in higher OF_{rob} limited by the corresponding BU and OF_{det} values. The obtained envelope bounds for every BU, presented in Table 8.3, are graphically compared in Fig. 8.14 to present a more precise evaluation with regard to the BU variation.

According to Fig. 8.14, by increasing the BU value, the growing trend is achieved for all envelope bounds of β_{lp}, β_{pv}, and β_{pr} leading to more extended size of the corresponding robust regions. The least differences of the envelope bounds are related to β_{lp} value between $BU = 0$ and $BU = 1$ as well as β_{pv} value between $BU = 1.25$ and $BU = 1.5$, as demonstrated in Fig. 8.14. Furthermore, the maximum differences all are associated with β_{pr} value between $BU = 0$ and $BU = 1.5$. It can be concluded from Fig. 8.14 that the alteration trend of most envelope bounds is proportionate to the maximum differences acquired by β_{pr} verifying the excellent robust performance of the proposed smart charging strategy. On the other hand, the proposed robust strategy is able to remain the robust purchase cost OF_{rob} within the increasing tendency as much as possible while the BU value is raised. This ability is due to the suitable performance of the MOPSO as well as the fuzzy satisfaction method through obtaining stable increasing values for all envelope bounds of β_{lp}, β_{pv}, and β_{pr}.

As another comparative discussion of Table 8.3, the difference between deterministic and robust purchase cost (i.e., $OF_{rob} - OF_{det}$) can be taken into consideration as a robustness criterion. In this way, the incremental behavior of the mentioned value justifies the robustness of the proposed charging model. As shown in Table 8.3, in the deterministic model where there is no uncertainty, all the envelope bounds of β_{lp}, β_{pv}, and β_{pr} are set to zero and so there is no immunity against the uncertainty realization leading to the lowest robustness. However, the proposed IGDT-based

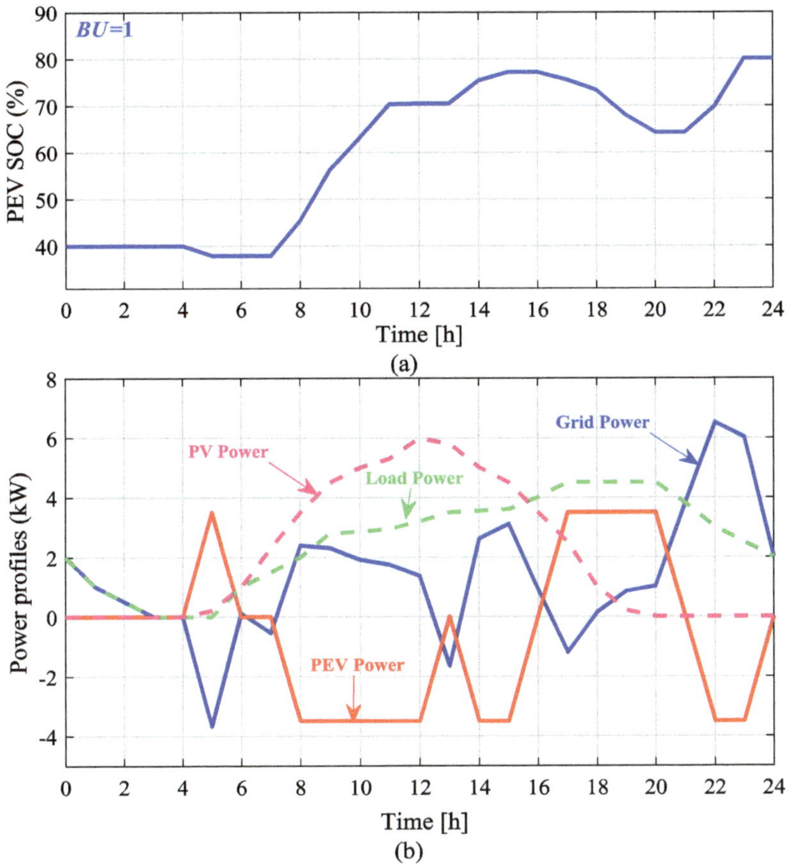

Fig. 8.9 The daily (**a**) SOC and (**b**) power profiles for $BU = 1$

charging model enhances the robustness of the decision-making framework by increasing the BU value leading to the higher envelope bounds of β_{lp}, β_{pv}, and β_{pr} at the expense of higher OF_{rob} value.

8.5 Conclusion

In this chapter, the main concentration is to achieve a smart charging strategy for a PEV integrated with exploiting the advantage of PV generation in a modern microgrid. Within the proposed EM-oriented model, the PEV user's viewpoint in terms of desired initial and final SOC pertaining to the PEV battery is incorporated. The attained results essentially rationalize the techno-economic benefits of the PEV in the contemporary microgrids. In this context, the specific advantage of the PEV to

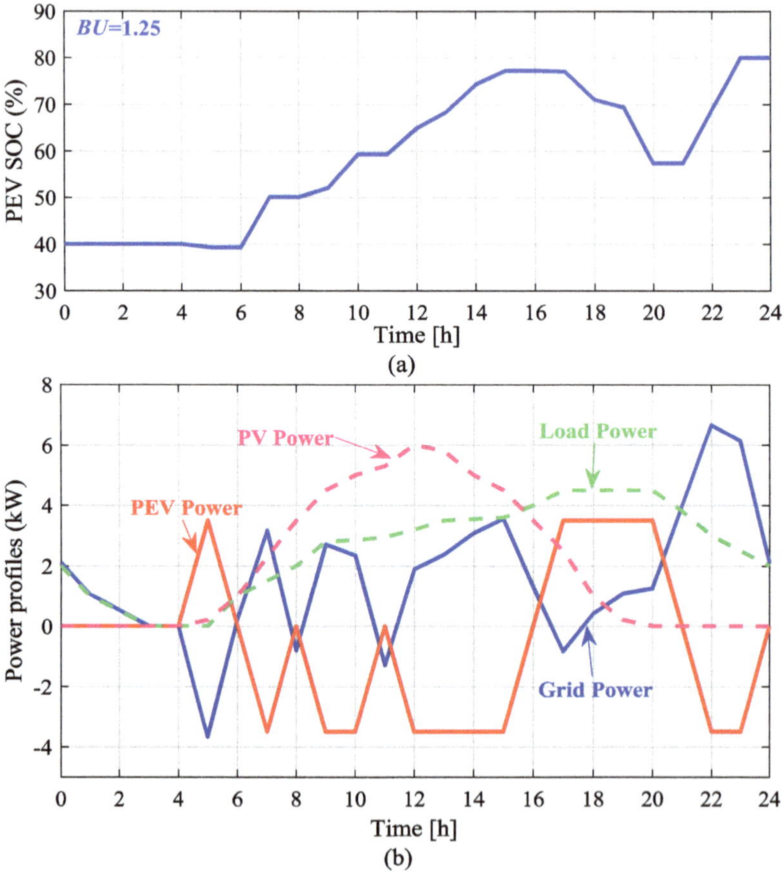

Fig. 8.10 The daily (**a**) SOC and (**b**) power profiles for $BU = 1.25$

be discharged in peak price time interval (generally concurrent with peak load time) is evidently observed. On the other hand, considering the potential contribution of the PV at the time of low electricity price, the PEV is optimally decided to be in either charge or idle mode. It is obvious that the PEV discharging mode helps to lessen the total cost. Moreover, although no power discharge is available in idle mode, this operation status can have a decremental effect on the PEV charging cost as a main portion of the total cost.

However, the main efficacious variables in the decision-making strategy of the presented model encounter severe uncertainties in practice such as load power, PV generation power, and also electricity price. Hence, to handle the mentioned uncertainty resources, the notion of information gap decision theory (IGDT) is utilized in this chapter. Using the IGDT, the decision outcomes of the proposed non-deterministic charging model are protected against the most probable deviation of uncertain variables from their anticipated values lying in their own robust regions.

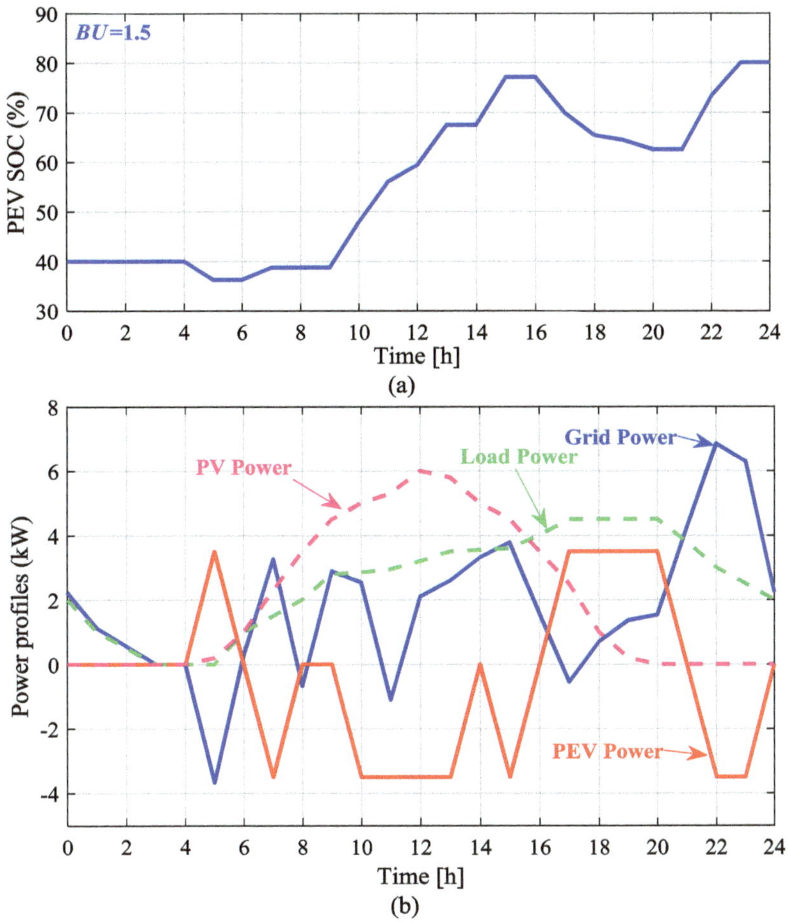

Fig. 8.11 The daily (**a**) SOC and (**b**) power profiles for $BU = 1.5$

In other words, the IGDT tends to maximize the robust regions of the uncertain variables wherein the uncertain variables can freely experience the different values rather than the corresponding desired values. Since any of the uncertain variables tries to maximize its own robust region, the proposed IGDT-based PEV charging model is fundamentally a multi-objective optimization problem. To solve such a problem, the multi-objective variant of particle swarm optimization (MOPSO) is used to provide a set of non-dominated Pareto solutions. Afterward, to reach the best compromise solution among the Pareto optimal solutions, a fuzzy satisfaction method is introduced and utilized. To assess the effectiveness of the proposed IGDT-based model as well as the MOPSO and fuzzy satisfaction method, a test microgrid including a set of PEV, PV unit, and electrical load is considered. The final solution is the most robust PEV charging strategy immunized against

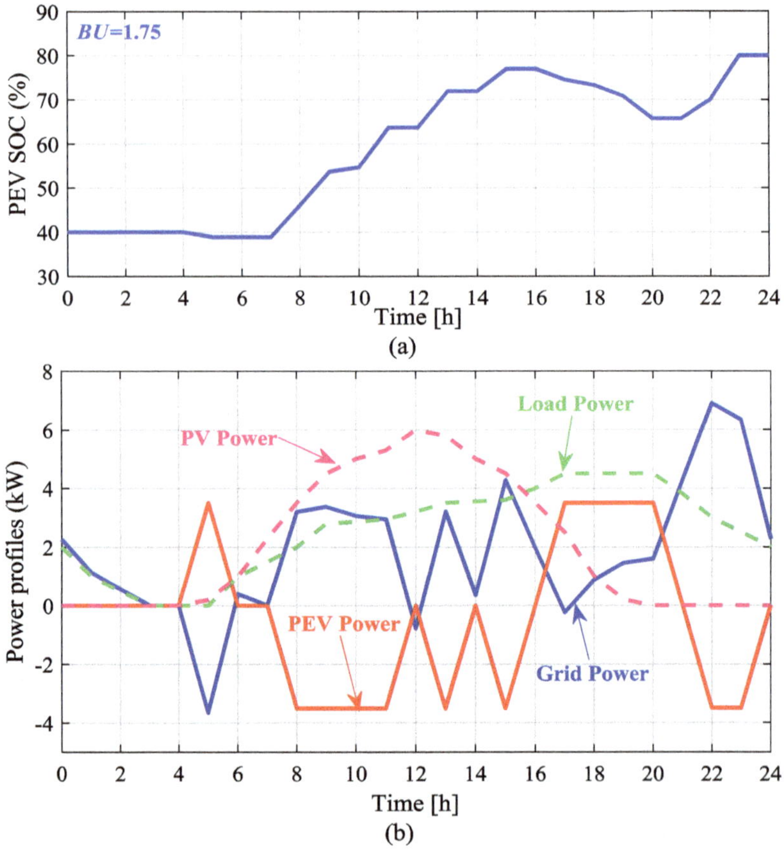

Fig. 8.12 The daily (**a**) SOC and (**b**) power profiles for $BU = 1.75$

any fluctuation of the uncertain variables within the associated robust regions. Additionally, the comparative study in terms of changes in the IGDT controllable parameter, called *BU*, demonstrates that the robust regions of the uncertain variables are more extended by an increase in *BU* value. In this circumstance, more robustness of the PEV charging pattern is attained at the expense of higher robust total cost (OF_{rob}) imposed to the decision-maker (grid operator/charging station owner in this chapter).

The proposed robust PEV smart charging strategy is structurally developed as a generic model and does not have any restriction to consider additional uncertainty resources (e.g., PEV arrival/departure time). Incorporating the mentioned issue into the proposed model of this chapter can be taken into account in the future work.

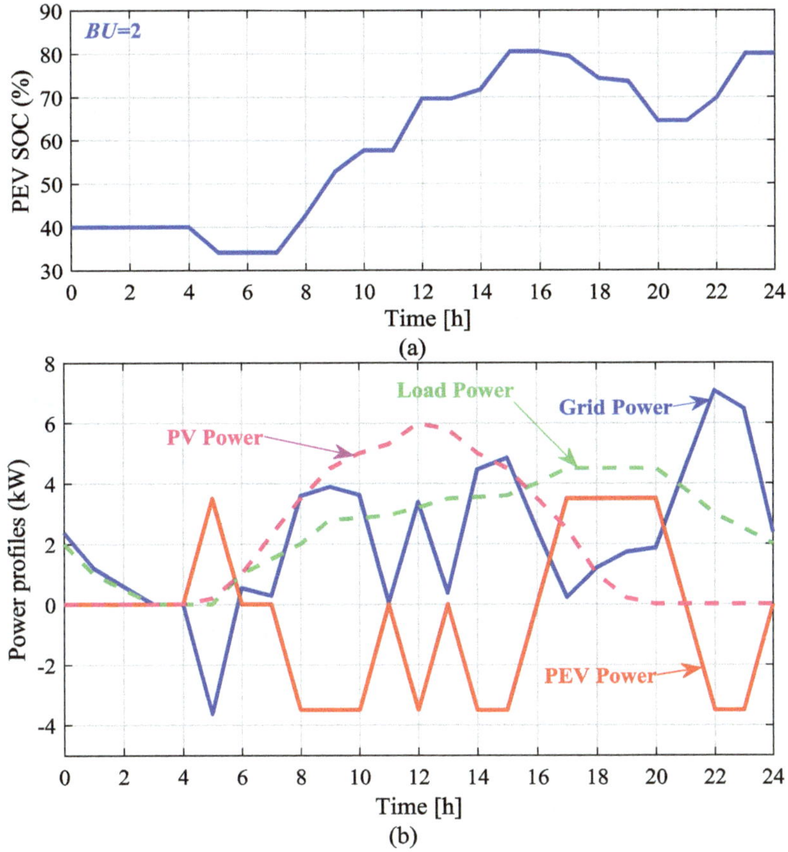

Fig. 8.13 The daily (**a**) SOC and (**b**) power profiles for $BU = 2$

Table 8.3 The numerical outputs associated with all BU values

BU	β_{lp}	β_{pv}	β_{pr}	OF_{rob} (S)	$OF_{rob} - OF_{det}$ (\$)
0.00	0.00	0.00	0.00	2.5605	0
1.00	0.0060	0.1091	0.1480	5.6704	1.4299
1.25	0.0539	0.1694	0.2738	7.3699	3.1294
1.50	0.1160	0.1704	0.3239	8.4777	4.6572
1.75	0.1339	0.2652	0.3870	9.3607	5.5402
2.00	0.1912	0.3453	0.4587	11.3940	7.5735

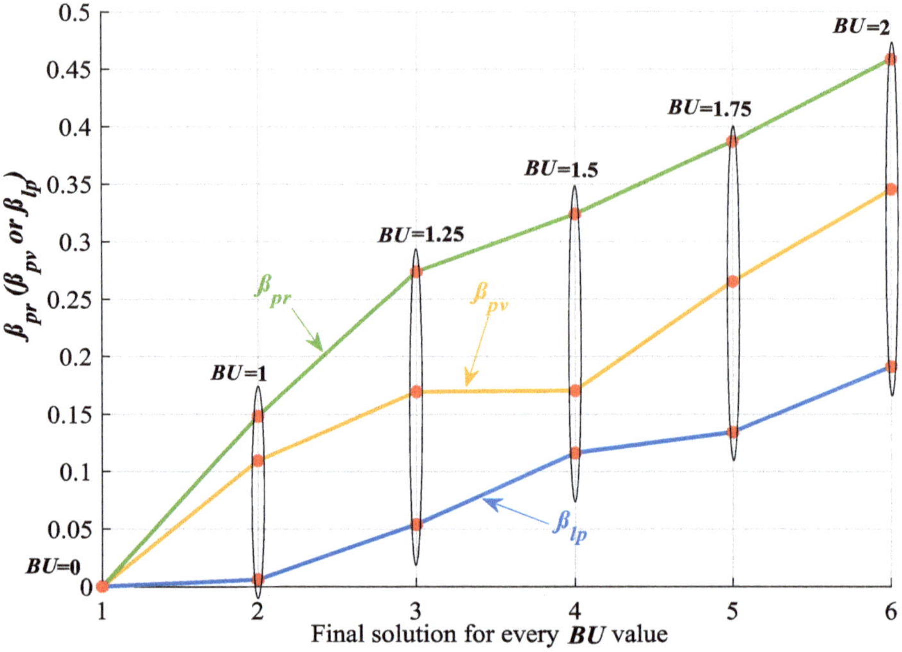

Fig. 8.14 The difference between β values attained for the final solution of every *BU* value

Acknowledgment This work has been supported by the aVEnir project of the PIA operated by ADEME.

References

1. Pahlavanhoseini, A., & Sepasian, M. S. (2019). Optimal planning of PEV fast charging stations using Nash bargaining theory. *Journal of Energy Storage, 25*, 1–14.
2. Wang, L., & Chen, B. (2019). Distributed control for large-scale plug-in electric vehicle charging with a consensus algorithm. *International Journal of Electrical Power & Energy Systems, 109*, 369–383.
3. Liang, H., Liu, Y., Li, F., & Shen, Y. (2019). Dynamic economic/emission dispatch including PEVs for peak shaving and valley filling. *IEEE Transactions on Industrial Electronics, 66*(4), 2880–2890.
4. Huang, Y. (2019). Day-ahead optimal control of PEV battery storage devices taking into account the voltage regulation of the residential power grid. *IEEE Transactions on Power Apparatus and Systems, 34*(6), 4154–4167.
5. Chakraborty, D., Hardman, S., & Tal, G. (2022). Integrating plug-in electric vehicles (PEVs) into household fleets- factors influencing miles traveled by PEV owners in California. *Travel Behavior Society, 26*, 67–83.
6. Sa'adati, R., Jafari-Nokandi, M., & Saebi, J. (2021). Allocation of RESs and PEV fast-charging station on coupled transportation and distribution networks. *Sustainable Cities and Society, 65*, 102527. https://doi.org/10.1016/j.scs.2020.102527

7. Sun, S., Yang, Q., & Yan, W. (2018). A novel Markov-based temporal-SoC analysis for characterizing PEV charging demand. *IEEE Transactions on Industrial Informatics, 14*(1), 156–166.
8. Patil, H., & Kalkhambkar, V. N. (2021). Grid integration of electric vehicles for economic benefits: A review. *Journal of Modern Power Systems and Clean Energy, 9*(1), 13–26.
9. Sangswang, A., & Konghirun, M. (2020). Optimal strategies in home energy management system integrating solar power, energy storage, and vehicle-to-grid for grid support and energy efficiency. *IEEE Transactions on Industry Applications, 56*(5), 5716–5728.
10. Singh, J., & Tiwari, R. (2020). Cost benefit analysis for V2G implementation of electric vehicles in distribution system. *IEEE Transactions on Industry Applications, 56*(5), 5963–5973.
11. Mehta, R., Srinivasan, D., Khambadkone, A. M., Yang, J., & Trivedi, A. (2018). Smart charging strategies for optimal integration of plug-in electric vehicles within existing distribution system infrastructure. *IEEE Transactions on Smart Grid, 9*(1), 299–312.
12. Shang, Y., Liu, M., Shao, Z., & Jian, L. (2020). Internet of smart charging points with photovoltaic Integration: A high-efficiency scheme enabling optimal dispatching between electric vehicles and power grids. *Applied Energy, 278*, 115640. https://doi.org/10.1016/j.apenergy.2020.115640
13. Turker, H., & Bacha, S. (2018). Optimal minimization of plug-in electric vehicle charging cost with vehicle-to-home and vehicle-to-grid concepts. *IEEE Transactions on Vehicular Technology, 67*(11), 10281–10292.
14. Ovalle, A., Hably, A., Bacha, S., Ramos, G., & Hossain, J. M. (2016). Escort evolutionary game dynamics approach for integral load management of electric vehicle fleets. *IEEE Transactions on Industrial Electronics, 64*(2), 1358–1369.
15. Qi, C., Zhu, Y., Song, C., Yan, G., Xiao, F., Wang, D., Zhang, X., Cao, J., & Song, S. (2022). Hierarchical reinforcement learning based energy management strategy for hybrid electric vehicle. *Energy, 238*, 121703. https://doi.org/10.1016/j.energy.2021.121703
16. Qi, J., Lai, C., Xu, B., Sun, Y., & Leung, K.-S. (2018). Collaborative energy management optimization toward a green energy local area network. *IEEE Transactions on Industrial Informatics, 14*(12), 5410–5418.
17. Kandil, S. M., Farag, H. E. Z., Shaaban, M. F., & ZakiEl-Sharafy, M. (2018). A combined resource allocation framework for PEVs charging stations, renewable energy resources and distributed energy storage systems. *Energy, 143*, 961–972.
18. Ovalle, A., Hably, A., & Bacha, S. (2018). *Grid optimal integration of electric vehicles: Examples with Matlab implementation*. Springer. ISSN 2198-4182.
19. Florescu, A., Bratcu, A. I., Munteanu, I., Rumeau, A., & Bacha, S. (2015). LQG optimal control applied to on-board energy management system of all-electric vehicles. *IEEE Transactions on Control Systems Technology, 23*(4), 1427–1439.
20. Teng, Y., Cao, Y., Liu, M., Richard Yu, F., & Leung, V. C. M. (2021). Efficient blockchain-enabled large scale parked vehicular computing with green energy supply. *IEEE Transactions on Vehicular Technology, 70*(9), 1427–1439.
21. Zhu, M., Liu, X.-Y., & Wang, X. (2018). Joint transportation and charging scheduling in public vehicle systems- A game theoretic approach. *IEEE Transactions on Intelligent Transportation Systems, 19*(8), 2407–2419.
22. Sadeghi, D., Hesami Naghshbandy, A., & Bahramara, S. (2020). Optimal sizing of hybrid renewable energy systems in presence of electric vehicles using multi-objective particle swarm optimization. *Energy, 209*, 118471. https://doi.org/10.1016/j.energy.2020.118471
23. Kobashi, T., Yoshida, T., Yamagata, Y., Naito, K., Pfenninger, S., Say, K., Takeda, Y., Ahl, A., Yarime, M., & Harak, K. (2020). On the potential of "Photovoltaics + Electric vehicles" for deep decarbonization of Kyoto's power systems: Techno-economic-social considerations. *Applied Energy, 275*, 115419. https://doi.org/10.1016/j.apenergy.2020.115419
24. Giordano, F., Ciocia, A., Di Leo, P., Mazza, A., Spertino, F., Tenconi, A., & Vaschetto, S. (2020). Vehicle-to-home usage scenarios for self-consumption improvement of a residential prosumer with photovoltaic roof. *IEEE Transactions on Industry Applications, 56*(3), 2945–2956.

25. Vermeer, W., Mouli, G. R. C., & Bauer, P. (2020). Real-time building smart charging system based on PV forecast and Li-Ion battery degradation. *Energies, 13*(13), 3415. https://doi.org/10.3390/en13133415
26. Fachrizal, R., & Munkhammar, J. (2020). Improved photovoltaic self-consumption in residential buildings with distributed and centralized smart charging of electric vehicles. *Energies, 13*(5), 1153. https://doi.org/10.3390/en13051153
27. Ahmad, J., Tahir, M., & Mazumder, S. K. (2019). Improved dynamic performance and hierarchical energy management of microgrids with energy routing. *IEEE Transactions on Industrial Informatics, 15*(6), 3218–3229.
28. Yousefi, M., Hajizadeh, A., & Soltani, M. (2018). Energy management strategies for smart home regarding uncertainties: State of the art, trends, and challenges. In *IEEE International Conference on Industrial Technology (ICIT)* (pp. 1219–1225).
29. Van Der Meer, D., Mouli, G. R. C., Mouli, G. M. E., Elizondo, L. R., & Bauer, P. (2018). Energy management system with PV power forecast to optimally charge EVs at the workplace. *IEEE Transactions on Industrial Informatics, 14*(1), 311–320.
30. Olatunde, O., Yusri Hassan, M., Abdullah, M. P., & Abdul Rahmana, H. (2020). Hybrid photovoltaic/small-hydropower microgrid in smart distribution network with grid isolated electric vehicle charging system. *Journal of Energy Storage, 31*, 101673. https://doi.org/10.1016/j.est.2020.101673
31. Szinai, J. K., Sheppard, C. J. R., Abhyankar, N., & Gopal, A. R. (2020). Reduced grid operating costs and renewable energy curtailment with electric vehicle charge management. *Energy Policy, 136*, 111051. https://doi.org/10.1016/j.enpol.2019.111051
32. Carli, R., Dotoli, M., Jantzen, J., Kristensen, M., & Othmand, S. B. (2020). Energy scheduling of a smart microgrid with shared photovoltaic panels and storage: The case of the Ballen marina in Samsø. *Energy, 198*, 117188. https://doi.org/10.1016/j.energy.2020.117188
33. Eldeeb, H. H., Faddel, S., & Mohammed, O. A. (2018). Multi-objective optimization technique for the operation of grid tied PV powered EV charging station. *Electric Power Systems Research, 164*, 201–211.
34. Li, Y., Ni, Z., Zhao, T., Zhong, T., Liu, Y., Wu, L., & Zhao, Y. (2020). Supply function game based energy management between electric vehicle charging stations and electricity distribution system considering quality of service. *IEEE Transactions on Industry Applications, 56*(5), 5932–5943.
35. Ioakimidis, C. S., Thomas, D., Rycerski, P., & Genikomsakis, K. N. (2018). Peak shaving and valley filling of power consumption profile in non-residential buildings using an electric vehicle parking lot. *Energy, 148*, 148–158.
36. Wang, Y., Yang, Z., Mourshed, M., Guo, Y., Niud, Q., & Zhua, X. (2019). Demand side management of plug-in electric vehicles and coordinated unit commitment: A novel parallel competitive swarm optimization method. *Energy Conversion and Management, 196*, 935–949.
37. Reddy, K., Meikandasivam, S., & Vijayakumar, D. (2019). A novel strategy for maximization of plug-in electric vehicle's storage utilization for grid support with consideration of customer flexibility. *Electric Power Systems Research, 170*, 158–175.
38. Jannesar, M. R., Sedighi, A., Savaghebi, M., Anvari-Moghaddam, A., & Guerrero, J. M. (2019). Optimal probabilistic planning of passive harmonic filters in distribution networks with high penetration of photovoltaic generation. *International Journal of Electrical Power & Energy Systems, 110*, 332–348.
39. Zhang, J., Xiong, G., Meng, K., Yu, P., Yao, G., & Dong, Z. (2019). An improved probabilistic load flow simulation method considering correlated stochastic variables. *International Journal of Electrical Power & Energy Systems, 111*, 260–268.
40. Sannigrahi, S., Ghatak, S. R., & Acharjee, P. (2020). Multi-scenario based bi-level coordinated planning of active distribution system under uncertain environment. *IEEE Transactions on Industry Applications, 56*(1), 850–863.
41. Günay, E. E., Kremer, G. E. O., & Zarindast, A. (2021). A multi-objective robust possibilistic programming approach to sustainable public transportation network design. *Fuzzy Sets and Systems, 422*, 106–129.

42. Tsao, Y.-C., Thanh, V.-V., & Lu, J.-C. (2019). Multiobjective robust fuzzy stochastic approach for sustainable smart grid design. *Energy, 176*, 929–939.
43. Zhou, Y., Li, Y. P., & Huang, G. H. (2015). A robust possibilistic mixed-integer programming method for planning municipal electric power systems. *International Journal of Electrical Power & Energy Systems, 73*, 757–772.
44. Ben-Haim, Y. (2006). *Information gap decision theory, design under severe uncertainty* (2nd ed.). Academic.
45. Ahmadigorji, M., Amjady, N., & Dehgan, S. (2018). A robust model for multiyear distribution network reinforcement planning based on information-gap decision theory. *IEEE Transactions on Power Apparatus and Systems, 32*(3), 1339–1351.
46. Rezaei, N., Ahmadi, A., Khazali, A., & Aghaei, J. (2019). Multiobjective risk-constrained optimal bidding strategy of smart microgrids: An IGDT-based normal boundary intersection approach. *IEEE Transactions on Industrial Informatics, 15*(3), 1532–1543.
47. Shafiee, S., Zareipour, H., Knight, A. M., Amjady, N., & Mohammadi-Ivatloo, B. (2017). Risk-constrained bidding and offering strategy for a merchant compressed air energy storage plant. *IEEE Transactions on Power Apparatus and Systems, 32*(2), 946–957.
48. Mansour-Saatloo, A., Ebadi, R., Mirzaei, M. A., Zare, K., Mohammadi Ivatloo, B., Marzband, M., & Anvari-Moghaddam, A. (2021). Multi-objective IGDT-based scheduling of low-carbon multi-energy microgrids integrated with hydrogen refueling stations and electric vehicle parking lots. *Sustain. Cities Society, 74*, 103197. https://doi.org/10.1016/j.scs.2021.103197
49. Nojavan, S., Majidi, M., & Zare, K. (2017). Risk-based optimal performance of a PV/fuel cell/battery/grid hybrid energy system using information gap decision theory in the presence of demand response program. *International Journal of Hydrogen Energy, 42*, 11857–11867.
50. Kennedy, J., & Eberhart, R. (1995). Particle swarm optimization. In *IEEE international conference on neural networks*.
51. Jallal, M. A., Chaba, S., & Zeroual, A. (2020). A novel deep neural network based on randomly occurring distributed delayed PSO algorithm for monitoring the energy produced by four dual-axis solar trackers. *Renewable Energy, 149*, 1182–1196.
52. Al-Bahrani, L. T., & Patra, J. C. (2018). Multi-gradient PSO algorithm for optimization of multimodal, discontinuous and non-convex fuel cost function of thermal generating units under various power constraints in smart power grid. *Energy, 147*, 1070–1091.
53. Abdelshafy, A. M., Hassan, H., & Juraszd, J. (2018). Optimal design of a grid-connected desalination plant powered by renewable energy resources using a hybrid PSO-GWO approach. *Energy Conversion and Management, 173*, 331–347.
54. Chaieb, H., & Sakly, A. (2018). A novel MPPT method for photovoltaic application under partial shaded conditions. *Solar Energy, 159*, 291–299.
55. Cheng, Y. S., Chuang, M. T., Liu, Y. H., Wang, S. C., & Yang, Z. Z. (2016). A particle swarm optimization based power dispatch algorithm with roulette wheel re-distribution mechanism for equality constraint. *Renewable Energy, 88*, 58–72.
56. Coello Coello, C. A., Pulido, G. T., & Lechuga, M. S. (2004). Handling multiple objectives with particle swarm optimization. *IEEE Transactions on Evolutionary Computation, 8*(3), 256–279.
57. Agrawal, R. K., & Bawane, N. G. (2015). Multiobjective PSO based adaption of neural network topology for pixel classification in satellite imagery. *Applied Soft Computing, 28*, 217–225.
58. Liu, X., Du, Y., Jiang, M., & Zeng, X. (2020). Multiobjective particle swarm optimization based on network embedding for complex network community detection. *IEEE Transactions on Computational Social Systems, 7*(2), 437–449.
59. Ahmadigorji, M., & Amjady, N. (2015). Optimal dynamic expansion planning of distribution systems considering non-renewable distributed generation using a new heuristic double-stage optimization solution approach. *Applied Energy, 156*, 655–665.
60. Soroudi, A., & Ehsan, M. (2013). IGDT based robust decision making tool for DNOs in load procurement under severe uncertainty. *IEEE Transactions on Smart Grid, 4*(2), 886–895.
61. Razi, R., Hajar, K., Hably, A., & Bacha, S. (2021, June). A user-friendly smart charging algorithm based on energy-awareness for different PEV parking scenarios. In *29th Mediterranean conference on control automation (MED)*.

Chapter 9
The Role of Smart Electric Vehicle Charging in Optimal Decision-making of the Active Distribution Network

Saeed Zeynali, Nima Nasiri, Sajad Najafi Ravadanegh, and Mousa Marzband

Nomenclature

Sets and Indices

t, s	Indices of time period, scenario
e, lk	Indices of EV fleets, linearization segments in battery degradation cost
h, p	Indices of CHP, PV
i, j	Indices of distribution system bus
l, d	Indices of interruptible load, distribution loads
A_i^p	Set of PVs connected to the bus i
A_i^h	Set of CHP units connected to the bus i
A_i^l	Set of interruptible load performed to the bus i
A_i^d	Set of electricity loads connected to the bus i
DS	Set of distribution network lines

S. Zeynali · N. Nasiri (✉) · S. N. Ravadanegh
Resilient Smart Grid Research Lab, Electrical Engineering Department, Azarbaijan Shahid Madani University, Tabriz, Iran

M. Marzband
Northumbria University, Electrical Power and Control Systems Research Group, Newcastle upon Tyne, UK

V. Vahidinasab, B. Mohammadi-Ivatloo (eds.), *Electric Vehicle Integration via Smart Charging*, Green Energy and Technology,
https://doi.org/10.1007/978-3-031-05909-4_9

Parameters

P_h^{Min}, P_h^{Max}	Minimum/maximum output electricity power of CHP (MW)
H_h^{Min}, H_h^{Max}	Minimum/maximum output heat power of CHP (MW)
P_h^A, P_h^B, P_h^C, P_h^D	Electricity power operational point of CHP (MW)
H_h^B, H_h^C	Heating power operational point of CHP (MW)
T_h^{Ue}/T_h^{De}	Up/down time of CHP (h)
R_h^{up}, R_h^{dn}	Up/down ramp rate of CHP (MW)
C_h^{SU}, C_h^{SD}	Startup/ shutdown fuel consumption of CHP (MW)
γ_{th}, γ_{el}	Conversion coefficients of heat and electrical power generated of CHP (%)
$P_{p,t}^{pvmax}$	Maximum output power of PV (MW)
R_{ij}^{DS}, Z_{ij}^{DS}	Resistance and impedance of distribution network feeders (Ω)
$I_{ij}^{DS\max}$, $I_{ij}^{DS\max}$	Maximum transit current of distribution network feeders (A)
$V_i^{DS\min}$, $V_i^{DS\max}$	Minimum/maximum voltage of distribution network bus (KV)
$P_{d,t}^{Load}$	Electricity demand of distribution network bus (MW)
π_s	Probability of scenario
η_c, η_e	Charge and discharge coefficients of EV fleets (%)
EB_e	Battery size of EV fleets
SOC_e^{Min}, SOC_e^{Max}	Minimum/maximum storage capacity of the EV fleets battery (KW)
$SOC_{e,s}^{In}$	Initial value of SOC in EV fleets (KW)
$ECPM_e$	Consumption of EV fleets energy per mile in fleet e (kWh/mile)
Cr_e	Nominal charging rate of fleet e (kWh)
a_0	Cycle depth degradation function coefficient
BC_e	Cost of EV fleets battery ($)
$\overline{y_{kl,e}^{aux1}}$, $\overline{y_{kl,e}^{aux2}}$	The value at the end of the linear segment for fleet *e* at linear segment *kl*, time *t*, and scenario *S*
$SOC_{e,s}^{end}$	Highest possible departure time SOC for fleet e and scenario s
SOC_e^{des}	The required SOC of EV when departure
Γ_0	Uncertainty budget
λ_t	Electricity power purchase price from upstream network($/MWh)

Variables

$SOC_{e,\,s,\,t}$	Amount of state of charge in the EV fleets (KW)
$PC_{e,\,s,\,t}$, $PD_{e,\,s,\,t}$	Amount of charge and discharge in the EV fleets (KW)

$\sigma_{e,s,t}$	Amount of cycle depth in the EV fleet
$\psi_{e,s,t}$	Amount of cycle depth degradation cost function of the EV fleet
$MD_{e,s,t}$	Amount of marginal battery degradation cost in the EV fleet ($/KWh)
$FY^{aux1}_{e,s,t}$, $FY^{aux2}_{e,s,t}$	Continuous auxiliary variables defined in the process of linearizing battery degradation of the fleet
$dg^{Lin}_{e,s,t}$	Linearized battery degradation of EV fleet
$dg^{NL}_{e,s,t}$	Nonlinear battery degradation of EV fleet
$P^{CHP}_{h,s,t}$, $H^{CHP}_{h,s,t}$	Amount of electricity and heat power generated of CHP units (MW)
$C^{chp}_{h,s,t}$	Amount of natural gas fuel required for CHP units (KCf)
$SU_{h,s,t}$, $SD_{h,s,t}$	Amount of natural gas fuel required for startup/shutdown CHP units (KCf)
$P^{pv}_{p,s,t}$	Amount of power generated by PV units (MW)
$P^{flow}_{ij,t}$	Amount of electricity power flow in the distribution network feeders (MW)
$P^{Loss}_{ij,t}$	Amount of electricity power losses in the distribution network feeders (MW)
$I^{DS}_{ij,t}$	Amount of transit current in the distribution network feeders (A)
$V^{DS}_{i,t}$	Amount of voltage in the distribution network buss (KV)
$P^{IL}_{l,t}$	Amount of interruptible load (MW)
P^{m}_{t}	Amount of power exchanged between DSO and TSO (MW)
$y^{RO}_{p,s,t}$, $p^{RO}_{p,s,t}$, $z^{RO}_{p,s,t}$	Auxiliary variables related to robust optimization

Binary Variables

$I_{h,t,s}$	Commitment state of CHP units
$y_{h,t,s}$, $z_{h,t,s}$	Startup and shutdown state of CHP units

9.1 Introduction

The increasing penetration of electric vehicles in distribution systems has given rise to numerous predicaments, such as voltage drops, high losses, and insecure operational states. These issues are exacerbated even more by uncertain production of the solar distributed generation units, electricity price, and unpredictability that is driven by behavioral patterns of electric vehicle (EV) owners. The researchers have illustrated that the peak demand of EVs (when they are charged uncoordinatedly) will co-occur with conventional residential peak demand, which can put a great deal of tension on the distribution systems. To account for these issues this book chapter investigated the coordinated scheduling of smart residential EV parking lots (SEVPL) in electric distribution systems, and solar distributed generation (SDG). The vehicles have been congregated into fleets, utilizing the K-means clustering, then the stochastic programming (SP) method was deployed to deal with indefinite parameters such as their arrival time, departure time, and daily traveled miles. Furthermore, to integrate a risk measure into the problem, robust optimization method (RO) has been deployed in parallel to SP for the purpose of dealing with uncertain parameters that do not follow any specific probability distribution, such as solar energy production and energy price. Additionally, the emerging combined heat and power (CHP) and non-gas-fired distributed generation technologies have been considered in the problem. Eventually, the book chapter investigated the impacts of coordinated and uncoordinated EV charging modes; and the benefits of utilizing the proposed RO-SP method are justified by various case studies.

9.2 Status Quo, Challenges, and Outlook

Many studies have been conducted on the optimal management of EVs in the presence of a variety of flexible energy sources. Optimal management of EVs in the presence of demand response program and electrical energy storage systems has been investigated in [1] and [2], respectively. Also, optimal management of EVs in coordination with CHP units and hydrogen storage systems has been studied in [3] and [4], respectively. The authors of ref. [5] have presented an optimal energy management approach for an EV intelligent parking in the presence of renewable energy source, distributed generation units, and hydrogen storage tank with the aim of minimizing the operation cost. The authors of [6] have investigated an optimal interval scheduling for a smart parking lot considering the demand response programming and the effect of the stochastic behavior of the electricity market price. An information gap decision theory model for optimal management of smart parking in an uncertainty environment has been studied in order to obtain the optimal proposed curve in [7]. The results were evaluated in three strategies: risk aversion, neutral risk, and risk seeker. An optimal management strategy for EV parking lots has been evaluated by considering demand response programming with the aim of reducing peak load under a stochastic approach in [8]. The authors of [9] have presented a bi-level optimization problem for participation of the EV aggregator in the wholesale market. In this research, the optimal scheduling of EV and clearing of electricity wholesale market has been modeled on the upper and lower of the problem, respectively. However, studies [1–9] ignore the impact of the high penetration of EVs on the distribution network decision. In this way, the focus of some other authors is on the high penetration of EVs in power distribution networks. The effect of high penetration of EVs on the stability of the electricity distribution network has been investigated in [10]. Also, the impact of EVs on increasing the flexibility and improving the resilience of the power distribution network has been evaluated in [11] and [12], respectively. In addition, the impact of EVs on increasing power quality and improving the reliability of smart electric distribution networks has been studied in [13] and [14], respectively. In [15], a stochastic optimization model is proposed to assess the impact of EVs penetration considering factors of vehicle class, battery capacity, vehicle charging mode, and EV charging cost in the electric distribution network. The authors of [16] have evaluated the impact of EVs penetration on a low voltage electric distribution network based on controlled and uncontrolled charging. The results show that, in a controlled charge, more EVs can be charged from the electric distribution network without upgrading the infrastructure. An optimal model for managing the active and reactive power of the smart electric distribution network with the aim of minimizing operating costs and improving the voltage profile by using the ability to control the active and reactive power of EVs conjugates in [17] is presented. In [18], a study is presented on assessing the reliability of the electric distribution network in the widespread presence of EVs, distributed generations, and photovoltaic units. In [19], a two-stage optimization problem is studied to investigate the effect of EVs and electric distribution network reconfiguring on losses and grid voltage profiles using a genetic algorithm to solve the problem. However, the literature [10–19] does not consider the impact of the uncertain behavior of wholesale energy market prices, the stochastic behavior of electric vehicle owners, and the uncertainty of PVs. For the gap of the above studies, a stochastic-robust hybrid optimization approach is proposed to evaluate the impact of high penetration of EVs in the electricity distribution network in the presence of CHP, non-gas fired distributed units, and PVs with the aim of minimizing total operating costs. Also, the impact of uncontrolled charging and smart charging in EVs on the distribution network decision will be evaluated. The general structure of the proposed approach is presented in Fig. 10.1. According to Fig. 9.1, the distribution company (Disco) first collects security data from the distribution network operator (DSO), the amount of price and generation capacity of the distributed units, the estimated generation capacity of PVs, the amount of estimated electricity demand, and the amount of power consumption of electric vehicles. After collecting information, the distribution company submits its offers/bids to the wholesale electricity market. The Disco uses a robust optimization approach to deal with the uncertain behavior of the wholesale electricity market price. In addition, stochastic optimization method has been used to model the uncertainty of PVs and EV owners.

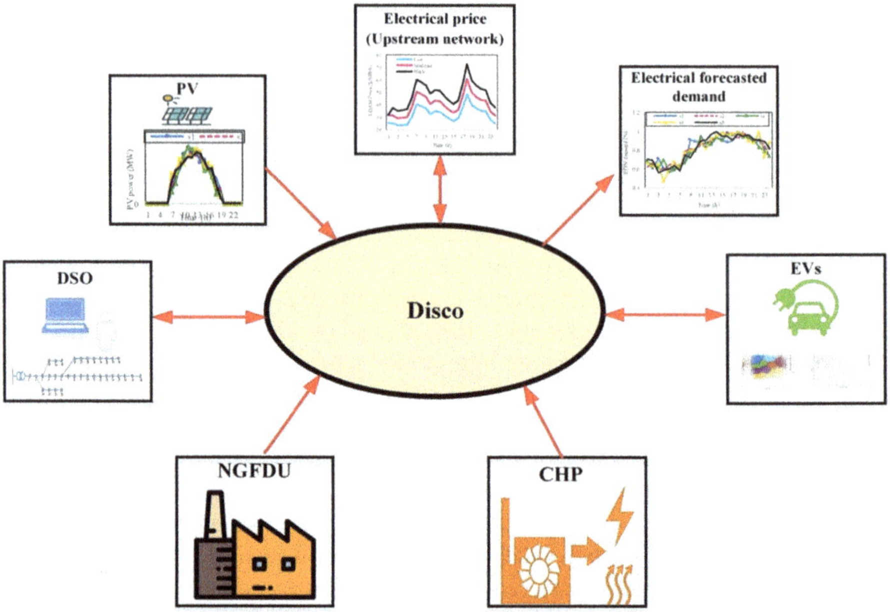

Fig. 9.1 General structure of the proposed approach

9.3 Formulation

9.3.1 Hybrid Stochastic Programming/Robust Optimization Model

Conventionally, a mixed-integer linear SP problem is expressed as follows:

$$\min \sum_{s} \pi_s \sum_{j \in WS} c'_{s,j} x_{s,j} + \sum_{j \in HN} c_j x_j \tag{9.1}$$

$$\sum_{j \in WS} A'_{s,j,i} x_{s,j} + \sum_{j \in HN} A_{j,i} x_j \leq b_{s,i} \tag{9.2}$$

$$\sum_{j \in WS} A'_{s,j,i} x_{s,j} + \sum_{j \in HN} A_{j,i} x_j = g_{s,i} \tag{9.3}$$

wherein s: index of scenario, WS: set of second stage SP variables, j: index of variables, i: index of constraints, HN: set of first stage variables, π_s: the probability of scenarios s, $c'_{s,j}$: uncertain parameters in the objective function, c_j: set of known parameters, $A'_{s,j,i}$: set of uncertain parameters in the constraint equations, $A_{j,i}$:

deterministic parameters in the constraints, and $b_{s,\,i}/g_{s,\,i}$: the uncertain parameters in inequality/equality equations.

Accordingly, (9.1) represents the optimization objective, while equality and inequality constraints are expressed via (9.2)–(9.3). As can be seen, SP has two types of variables, which are namely first and second stage variables. The first stage corresponds to the decision variables that ought to be determined prior to the scheduling. However, the second stage variables are entangled with stochastic scenarios. The downside of SP is that its mean-oriented, and only the expected value of the fitness function is minimized, which in turn might lead to some very high-cost scenarios that are usually result from the stochastic scenarios that are at the edges of the probability distribution function. That said, robust optimization (RO) is known to provide reliable outcomes even in the worst scenario. Furthermore, it does not rely on probability distribution functions, which is a handy feature when these functions are either non-existent or hard to attain. Nevertheless, RO can sometimes be too conservative in the provided solutions that might not be a favorable attribute. In this regard, this study puts forward the following hybridized RO-SP formulation that eliminates over-conservatism and risky solutions problems that are associated with RO and SP models. Overall this model is formulated as follows:

$$\min \sum_{s} \pi_s \left(\sum_{j \in WS} c'_{s,j} x_{s,j} + \sum_{j \in HN} c_j x_j + z_0 \Gamma_0 + \sum_{j = J_0} \phi_{s,0,j} \right) \tag{9.4}$$

$$z_0 + \phi_{s,0,j} \geq D_j \Upsilon_{s,j} \;\; \forall s,\, j \in J_0 \tag{9.5}$$

$$z_i + \phi_{s,j,i} \geq \hat{A}_{j,i} \Upsilon_{s.i} \;\; \forall s,\, j \in J_i,\, i \neq 0 \tag{9.6}$$

$$-\Upsilon_{s,j} \leq x_{s,j} \leq \Upsilon_{s,j} \;\; \forall s,\, j \tag{9.7}$$

$$L_j \leq x_{s,j} \leq U_j \;\; \forall s,\, j \tag{9.8}$$

$$\phi_{s,j} \geq 0,\, \Upsilon_{s,j} \geq 0,\, z_{s,i} \geq 0 \;\; \forall s,\, i \tag{9.9}$$

wherein $z_0/\phi_{s,\,0,\,j}/z_i/\phi_{s,\,j,\,i}/\Upsilon_{s,\,j}$: auxiliary variables of robust optimization, Γ_0/Γ_i: robustness controller in the cost function/constraints, respectively, $D_j/\hat{A}_{j,i}$: forecasted deviation of the parameters from mean value, and L_j/U_j: the upper/lower variable bounds. Additionally, the strong duality theory is deployed to prove this model as it is elaborated in [20].

9.3.2 Electric Vehicles

In this study, SEVPLs are modeled on basis of the formulation presented in [9]. First of all, electric vehicles are congregated into distinct fleets according to the behavioral distributions of their owners by k-means clustering. The each electric vehicle fleet (EVF) is attributed to a set of stochastic scenarios that are generated according to their daily traveled miles and arrival/departure times. The SEVPLs are modeled by (9.10)–(9.21). The state of charge (SOC) of EVFs in each time interval is established by (9.10)–(9.11) for all time steps and arrival time, respectively. The initial SOC of each EVF is defined by (9.12), while the finial SOC at departure is expressed as the minimum between the desired SOC and highest feasible SOC (which can be obtained by charging vehicles nonstop from arrival time). Furthermore, the bounds of SOC, charge, and discharge powers are included by (9.14)–(9.16). The overlapping of charge/discharge events is avoided via the binary equation in (9.17). The cycle depth and cycle depth battery degradation function for EVFs is defined by (9.18)–(9.19), while (9.20) shows the way in which the marginal (per kW) battery degradation is attained. Eventually, the overall battery erosion is evaluated via (9.21) [21].

$$SOC_{e,s,t} = SOC_{e,s,t-1} + \frac{\eta_c . PC_{e,s,t}}{EB_e} - \frac{PD_{e,s,t}}{EB_e . \eta_e} \qquad \forall e, s, t : \{t \mid t \in t \neq TA_e\} \tag{9.10}$$

$$SOC_{e,s,t} = SOC^{In}_{e,s} + \frac{\eta_c . PC_{e,s,t}}{EB_e} - \frac{PD_{e,s,t}}{EB_e . \eta_e} \qquad \forall e, s, t : \{t \mid t \in t = TA_e\} \tag{9.11}$$

$$SOC^{In}_{e,s} = \max\left(SOC^{Min}_e, 1 - \frac{DT_{e,s} \times ECPM_e}{EB_e}\right) \qquad \forall e, s \tag{9.12}$$

$$SOC_{e,s,t} = \min\left(SOC^{end}_{e,s}, SOC^{des}_e\right) \qquad \forall e, s, t : \{t \mid t \in t = TD_e\} \tag{9.13}$$

$$SOC^{Min}_e \leq SOC_{e,s,t} \leq SOC^{Max}_e \qquad \forall e, s, t \tag{9.14}$$

$$PC_{e,s,t} \leq Cr_e . uc_{e,s,t} \qquad \forall e, s, t \tag{9.15}$$

$$PD_{e,s,t} \leq Cr_e . ud_{e,s,t} \qquad \forall e, s, t \tag{9.16}$$

$$uc_{e,s,t} + ud_{e,s,t} = 1 \qquad \forall e, s, t \tag{9.17}$$

$$\sigma_{e,s,t} = \frac{PD_{e,s,t}}{EB_e.\eta_e} - \sigma_{e,s,t-1} \qquad \forall e, s, t \tag{9.18}$$

$$\psi_{e,s,t}\left(\sigma_{e,s,t}\right) = a_0.\left(\sigma_{e,s,t}\right)^{2.03} \qquad \forall e, s, t \tag{9.19}$$

$$\begin{aligned} MD_{e,s,t} &= BC_e.\tfrac{\partial \psi_{e,s,t}}{\partial PD_{e,s,t}} = BC_e.\tfrac{\partial \psi_{e,s,t}}{\sigma_{e,s,t}}.\tfrac{\sigma_{e,s,t}}{\partial PD_{e,s,t}} \\ &= 2.03a_0.\tfrac{BC_e}{EB_e.\eta_e}.\sigma_{e,s,t}{}^{1.03} \end{aligned} \qquad \forall e, s, t \tag{9.20}$$

$$dg_{e,s,t}^{NL} = PD_{e,s,t}.MD_{e,s,t} \qquad \forall e, s, t \tag{9.21}$$

As can be observed, the production of the continuous variables in (9.21) makes it a nonlinear mathematical term. In this study, semi-integer variable linearization model is deployed to linearize this expression as it is illustrated in [22]. In this context, the special semi-integer variables are expressed by (9.22)–(9.23), which are then used to decompose (9.21) into a quadratic equation as it is shown by (9.24). This quadratic equation is linearized as it is illustrated by (9.25)–(9.28). In this model, $\hbar_{kl,e,s,t}^{a1}$ is a type two specially ordered semi-integer variable (SOS2) [23], which means that only two consecutive variables in a set can take non-zero values.

$$y_{e,s,t}^{aux1} = 0.5\left(PD_{e,s,t} + MD_{e,s,t}\right) \qquad \forall e, s, t \tag{9.22}$$

$$y_{e,s,t}^{aux2} = 0.5\left(PD_{e,s,t} - MD_{e,s,t}\right) \qquad \forall e, s, t \tag{9.23}$$

$$dg_{e,s,t}^{NL} = PD_{e,s,t}.MD_{e,s,t} = \left(y_{e,s,t}^{aux1}\right)^2 - \left(y_{e,s,t}^{aux2}\right)^2 \qquad \forall e, s, t \tag{9.24}$$

$$\begin{aligned} FY_{e,s,t}^{aux1} &= \sum_{kl=1}^{KL} \hbar_{kl,e,s,t}^{a1}\left(\overline{y_{kl,e}^{aux1}}\right)^2, \\ FY_{e,s,t}^{aux2} &= \sum_{kl=1}^{KL} \hbar_{kl,e,s,t}^{a2}\left(\overline{y_{kl,e}^{aux2}}\right)^2 \quad \forall e, s, t \end{aligned} \qquad \forall e, s, t \tag{9.25}$$

$$\begin{aligned} y_{e,s,t}^{aux1} &= \sum_{kl=1}^{KL} \hbar_{kl,e,s,t}^{a1}\left(\overline{y_{kl,e}^{aux1}}\right), \\ y_{e,s,t}^{aux2} &= \sum_{kl=1}^{KL} \hbar_{kl,e,s,t}^{a2}\left(\overline{y_{kl,e}^{aux2}}\right) \quad \forall e, s, t \end{aligned} \qquad \forall e, s, t \tag{9.26}$$

$$\sum_{kl=1}^{KL} \hbar^{a1}_{kl,e,s,t} = 1,\ \sum_{kl=1}^{KL} \hbar^{a2}_{kl,e,s,t} = 1\ \forall e, s, t \qquad \hbar^{a1}_{kl,e,s,t} > 0,\ \hbar^{a2}_{kl,e,s,t} > 0 \quad \forall e, s, t \tag{9.27}$$

$$dg^{Lin}_{e,s,t} = FY^{aux1}_{e,s,t} - FY^{aux2}_{e,s,t}\ \forall e, s, t \quad \forall e, s, t \tag{9.28}$$

9.3.3 Combined Heat and Power Unit

Combined heat and power units (CHPs) are high-efficiency co-generation technologies that generate electricity and thermal energy simultaneously with efficiency values that can be as high as 85%. The commitment status of the CHP units is declared by (9.29)–(9.30), which also limits their thermal electrical throughput. The feasible operation region limits of the CHPs are declared by (9.31)–(9.33). The ramp rate of the CHP is circumscribed via (9.34)–(9.35), while the start-up/shutdown modes are clarified by (9.36)–(9.37), which are binary equations. The limit on minimum turn-on time of the CHPs is declared by (9.38)–(9.41). Likewise, the minimum shutdown time limit is included in (9.42)–(9.45).

$$P^{Min}_h I_{h,s,t} \le P^{CHP}_{h,s,t} \le P^{Max}_h I_{h,s,t} \quad \forall h, \forall s, \forall t \tag{9.29}$$

$$0 \le H^{CHP}_{h,s,t} \le H^A_h I_{h,s,t} \quad \forall h, \forall s, \forall t \tag{9.30}$$

$$P^{CHP}_{h,s,t} - P^A_h - \frac{P^A_h - P^B_h}{H^A_h - H^B_h}\left(H^{CHP}_{h,t,s} - H^A_h\right) \le 0 \quad \forall h, \forall s, \forall t \tag{9.31}$$

$$P^{CHP}_{h,s,t} - P^B_h - \frac{P^B_h - P^C_h}{H^B_h - H^C_h}\left(H^{CHP}_{h,s,t} - H^B_h\right) \ge -\left(1 - I_{h,t,s}\right) M \quad \forall h, \forall s, \forall t \tag{9.32}$$

$$P^{CHP}_{h,s,t} - P^C_i - \frac{P^C_h - P^D_h}{H^C_h - H^D_h}\left(H^{CHP}_{h,s,t} - H^C_h\right) \ge -\left(1 - I_{h,t,s}\right) M \quad \forall h, \forall s, \forall t \tag{9.33}$$

$$P^{CHP}_{h,s,t} - P^{CHP}_{h,s,t-1} \le \left[1 - Y_{h,t,s}\right] R^{up}_h + Y_{h,t,s} P^{\min}_h \quad \forall h, \forall s, \forall t \tag{9.34}$$

$$P^{CHP}_{h,s,t-1} - P^{CHP}_{h,s,t} \le \left[1 - Z_{h,s,t}\right] R^{dn}_h + Z_{h,s,t} P^{\min}_h \quad \forall h, \forall s, \forall t \tag{9.35}$$

$$Y_{h,t} - Z_{h,t} = I_{h,t-1} - I_{h,t} \qquad \forall h, \forall s, \forall t \tag{9.36}$$

$$Y_{h,t} + Z_{h,t} \leq 1 \qquad \forall h, \forall s, \forall t \tag{9.37}$$

$$T_h^{Ue} = \min\left\{T, \left(T^{Ue} - T_h^{U0}\right) I_{h,t=0,s}\right\} \tag{9.38}$$

$$\sum_{t=1}^{T_i^{Ue}} I_{h,t} = T_h^{Ue} \qquad \forall h \tag{9.39}$$

$$\sum_{t=r}^{t+T_h^{Ue}-1} I_{h,r} \geq T_h^{Ue} y_{h,t} \qquad \forall h, \forall t = \left[T_h^{Ue} + 1, \ldots, T - T_h^{U} + 1\right] \tag{9.40}$$

$$\sum_{t=r}^{T} \left(I_{h,r} - y_{h,t}\right) \geq 0 \qquad \forall h, \forall t = \left[T - T_h^{Ue} + 2, \ldots\ldots, T\right] \tag{9.41}$$

$$T_i^{De} = \min\left\{T, \left(T^{De} - T_h^{U0}\right)\left(1 - I_{h,t=0,s}\right)\right\} \tag{9.42}$$

$$\sum_{t=1}^{T_h^{De}} I_{h,t} = 0 \quad \forall h \tag{9.43}$$

$$\sum_{t=r}^{t+T_h^{De}-1} \left(1 - I_{h,r}\right) \geq T_h^{De} z_{h,t} \qquad \forall h, \forall t = \left[T_h^{De} + 1, \ldots, T - T_h^{D} + 1\right] \tag{9.44}$$

$$\sum_{t=r}^{T} \left(1 - I_{h,r} - z_{h,t}\right) \geq 0 \qquad \forall h, \forall t = \left[T - T_h^{De} + 2, \ldots, T\right] \tag{9.45}$$

$$SU_{h,s,t} \geq C_h^{SU} Y_{h,s,t} \qquad \forall h, \forall s, \forall t \tag{9.46}$$

$$SD_{h,s,t} \geq C_h^{SD} Z_{h,s,t} \qquad \forall h, \forall s, \forall t \tag{9.47}$$

$$C_{h,s,t}^{chp} = H_{h,s,t}^{CHP} \cdot \gamma_{th} + P_{h,s,t}^{CHP} \cdot \gamma_{el} \qquad \forall h, \forall s, \forall t \tag{9.48}$$

9.3.4 Solar Distributed Generations

The robust solar generation model is established by (9.49)–(9.53). The min/max solar energy output of the PV systems is restricted by (9.49), while the robust solar generation model is established by (9.50)–(9.53), which is comprehensively described in Sect. 2.1.

$$0 \leq P_{p,s,t}^{pv} \leq P_{p,t}^{pvmax} \qquad \forall p, \forall s, \forall t \tag{9.49}$$

$$P_{p,s,t}^{pv} - x_{p,s,t}^{RO} P_{p,t}^{pv\max} + z_{p,s,t}^{RO} \Gamma^{RO} + p_{p,s,t}^{RO} \leq 0 \qquad \forall p, \forall s, \forall t \tag{9.50}$$

$$z_{p,s,t}^{RO} + p_{p,s,t}^{RO} \geq P_{p,s,t}^{pvDev} y_{p,s,t}^{RO} \qquad \forall p, \forall s, \forall t \tag{9.51}$$

$$-y_{p,s,t}^{RO} \leq x_{p,s,t}^{RO} \leq y_{p,s,t}^{RO} \qquad \forall p, \forall s, \forall t \tag{9.52}$$

$$y_{p,s,t}^{RO}, p_{p,s,t}^{RO}, z_{p,s,t}^{RO} > 0 \quad x_{p,s,t}^{RO} = 1 \qquad \forall p, \forall s, \forall t \tag{9.53}$$

9.3.5 Distribution System

The active distribution system (ADS) model is taken from [24]. The power flow in the feeders is determined by (9.54), while line delivery loss is calculated by (9.55). The current passing through the feeders can be obtained by (9.56), while the voltage and current levels are limited to their nominal amplitude by (9.57)–(9.59). The power equilibrium of slack and non-slack nodes is satisfied by (9.60)–(9.61).

$$P_{ij,t}^{flow} = \frac{R_{ij}^{DS}}{\left(Z_{ij}^{DS}\right)^2} \left(V_{i,t}^{DSsqr} - V_{j,t}^{DSsqr}\right) \qquad \forall ij, \forall t \tag{9.54}$$

$$P_{ij,t}^{Loss} = R_{ij}^{DS} I_{ij}^{DSsqr} \quad \forall ij, \forall t \tag{9.55}$$

$$I_{ij,t}^{DS} = \frac{V_{i,t}^{DS} - V_{j,t}^{DS}}{Z_{ij}^{DS}} \quad \forall ij, \forall t \tag{9.56}$$

$$-I_{ij}^{DS\max} \leq I_{ij,t}^{DS} \leq I_{ij}^{DS\max} \quad \forall ij, \forall t \tag{9.57}$$

$$V_i^{DS\min} \leq V_{i,t}^{DS} \leq V_i^{DS\max} \quad \forall i, \forall t \tag{9.58}$$

$$v_{gp}^{Min} \leq v_{gp,t} \leq v_{gp}^{Max} \quad \forall gp, \forall t \tag{9.59}$$

$$\begin{aligned} &P_t^m + \sum_{p \in A_i^p} P_{p,t}^{PV} + \sum_{h \in A_i^h} P_{h,t}^{CHP} + \sum_{l \in A_i^l} P_{l,t}^{IL} \\ &= \sum_{d \in A_i^d} P_{d,t}^{Load} + 0.5 \left(\sum_{j \in DS} P_{ij,t}^{Loss} + \sum_{j \in DS} P_{ij,t}^{flow} \right) \quad \forall i = 1, \forall t \end{aligned} \tag{9.60}$$

$$\begin{aligned} &\sum_{p \in A_i^p} P_{p,t}^{PV} + \sum_{h \in A_i^h} P_{h,t}^{CHP} + \sum_{l \in A_i^l} P_{l,t}^{IL} \\ &= \sum_{d \in A_i^d} P_{d,t}^{Load} + 0.5 \left(\sum_{j \in DS} P_{ij,t}^{Loss} + \sum_{j \in DS} P_{ij,t}^{flow} \right) \quad \forall i \neq 1, \forall t \end{aligned} \tag{9.61}$$

9.3.6 The Objective Function

The robust objective function is represented by (9.62), which is obtained considering the formulation in [20]. The first term in the bracket corresponds to the cost of operating CHPs, the second term is the renewable energy cost, and the third one is the battery degradation cost which is incurred on EVFs. Eventually, the last two statements correspond to the auxiliary variables that convert the objective function into a robust model, which are accompanied by complementary constraints in (9.63)–(9.67). Further information on the mathematical proof can be observed in [25].

$$\sum_s \pi_s \sum_t P_t^m \lambda_t \sum_h \left(C_{h,s,t}^{chp} + SU_{h,s,t} + SD_{h,s,t} \right) + \sum_e dg_{e,s,t}^{Lin} + \psi.\Gamma + \sum_t \zeta_t \tag{9.62}$$

$$\psi + \zeta_{t,s} \leq \left(\lambda_t^{Max} - \lambda_t^{Min} \right) y_{t,s} \qquad \forall t \tag{9.63}$$

$$\zeta_{t,s} \geq 0 \quad \forall t \tag{9.64}$$

$$y_{t,s} \geq 0 \qquad \forall t \tag{9.65}$$

$$P_{t,s}^{in} \leq y_{t,s} \qquad \forall t \tag{9.66}$$

$$\psi \geq 0 \tag{9.67}$$

9.4 Results and Discussions

The proposed ADS and the hypothetical location of the equipment and SEVPL is demonstrated in Fig. 9.2. It should be noted that nodes 1,5,9,10,11,13,18,31, and 32 are assumed to have interruptible loads (IL). Further, the dotted distribution of each EVFs' arrival-departure time can be observed in Fig. 9.3, while Fig. 9.4 depicts the empirical distribution of daily driven miles by each EVF. The data on ADS is taken from [26], while EVF data is provided in [9, 27]. It should be reiterated that EVs are divided into the clusters by K-means method. K-means clustering is applied to categorize EV user behavior and built for an online real-time EV user model to describe the uncertainty and eliminate the needs for hand labeling user behavior or revisiting historical dataset [28]. The proposed RO-SP model is deployed to handle uncertain entries. For instance, the solar irradiation is dealt with via the RO model, while EV-related arrival, departure, and driven distance are modeled by SP scenarios. The experiments are performed on a personal computer using a GAMS software in CPLEX with Intel Core (TM) i3-2348M CPU (2.3GHz) and 4 Gb of memory. Also, the average time to solve the proposed problem is 5 minutes for the first case study, 8 minutes for the second case study, and 10 minutes for the third case study.

To show the practicality of the smart EV scheduling and the proposed RO-SP method, following case studies are devised:

Case 1: The EVFs are charged uncoordinatedly as soon as they reach SEVPL using SP.

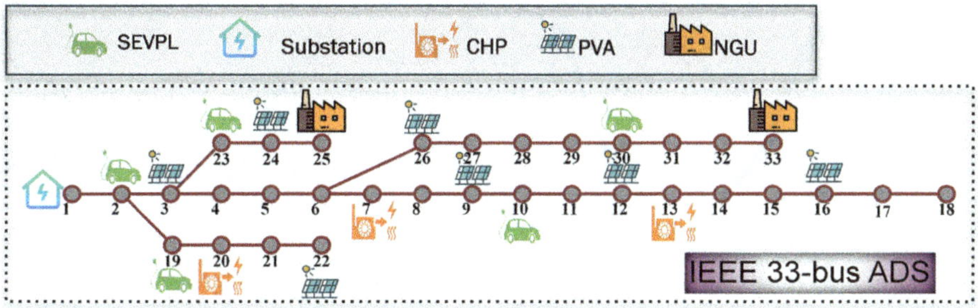

Fig. 9.2 The overall structure of the proposed system

Fig. 9.3 Arrival/departure distribution of EVFs

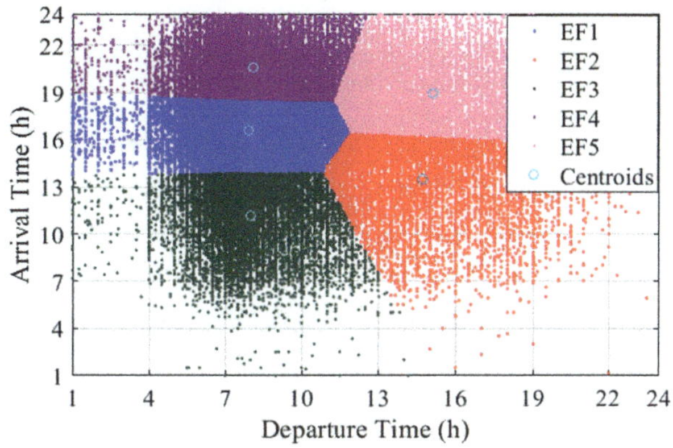

Fig. 9.4 Probability distribution of EVFs' daily driven distance

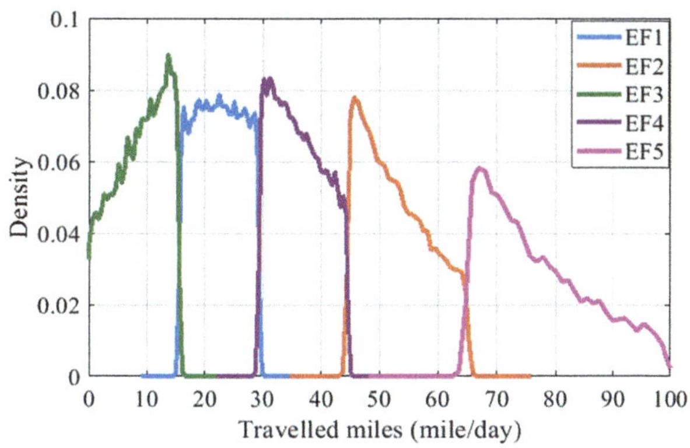

Case 2: The EVFs are charged smartly from their arrival to departure intervals using SP.

Case 3: The EVFs are charged smartly from their arrival to departure intervals using RO-SP. ($\Gamma = 24$)

The SOC of each EVF through different cases is plotted in Fig. 9.5, which happens at times

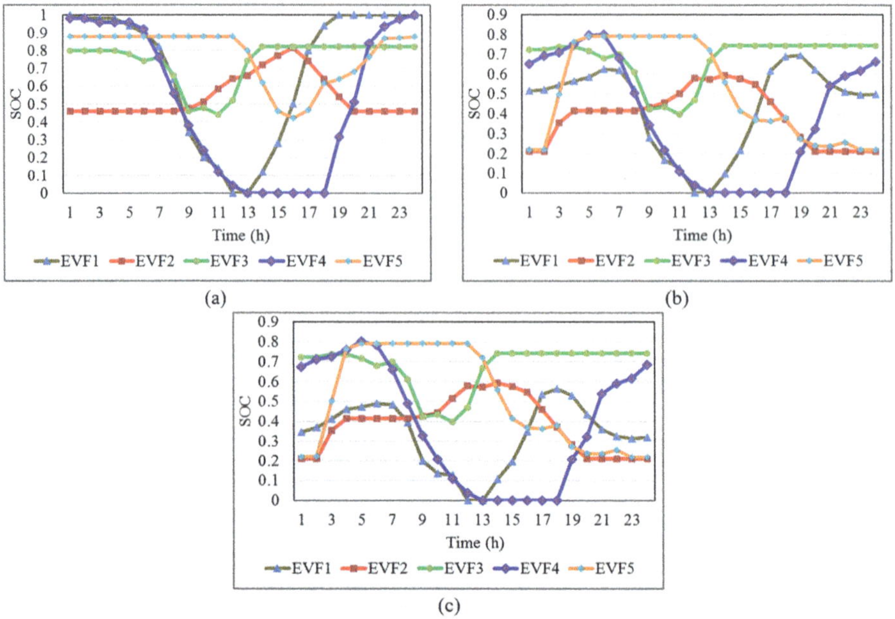

Fig. 9.5 The optimal schedule of all EVFs: (**a**) Case 1. (**b**) Case 2. (**c**) Case 3

As can be observed from Fig. 9.5, in Case 1 the SOC of all fleets demonstrates a sharp increase as soon as they reach SEVPL, which happens around hours 17–23. However, with smartly scheduled framework in Case 2, the charge periods are delayed back to hours 3–7, which is the least loaded period in the whole schedule span. Nevertheless, these variations are even more prominent in Case 3 since the market price and solar generation are predicted prudently to avoid risk. As a result, EVFs not only show more dramatic shift in their charge hours, but also slight discharge sequences are visible at peak hours, which are aimed at boosting the operation.

The electrical power dispatch of the units in various cases can be observed in Fig. 9.6.

As can be seen in Fig. 9.6, in Case 1 the peak EV load coincides with the residential demand peak. Therefore, this case imposes the highest dispatch at peak hours and even some of the ILs are interrupted to enforce security constrains. However, in Case 2 the scheduling is conducted smartly, which has successfully shifted the EV schedule to off-peak segments, thereby not only the load interruption is almost zero, but also the cheaper energy is purchased from the market during off-peak intervals. Additionally, less power is dispatched from expensive units. In case 3, the scheduling results are illustrated for the most robust scheduling case ($\Gamma = 24$). The most notable nuance between Case 3 and Case 2 is that not more of EV demand is shifted to earlier hours, but also EVFs are discharged at alarming peak intervals (pink color in Fig. 9.6c at hours 19,20, and 21). The obvious reason behind this

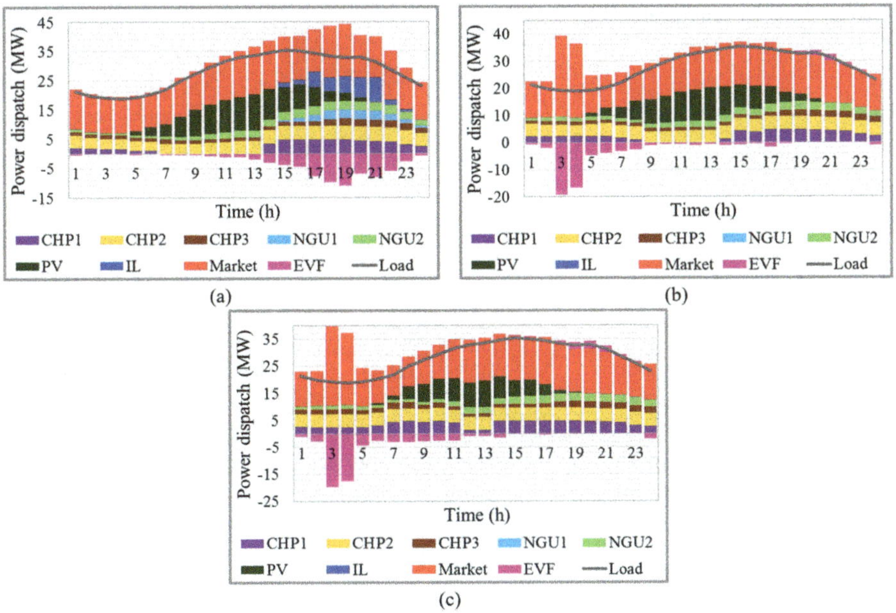

Fig. 9.6 The optimal schedule of all units and EVs: (**a**) Case 1. (**b**) Case 2. (**c**) Case 3

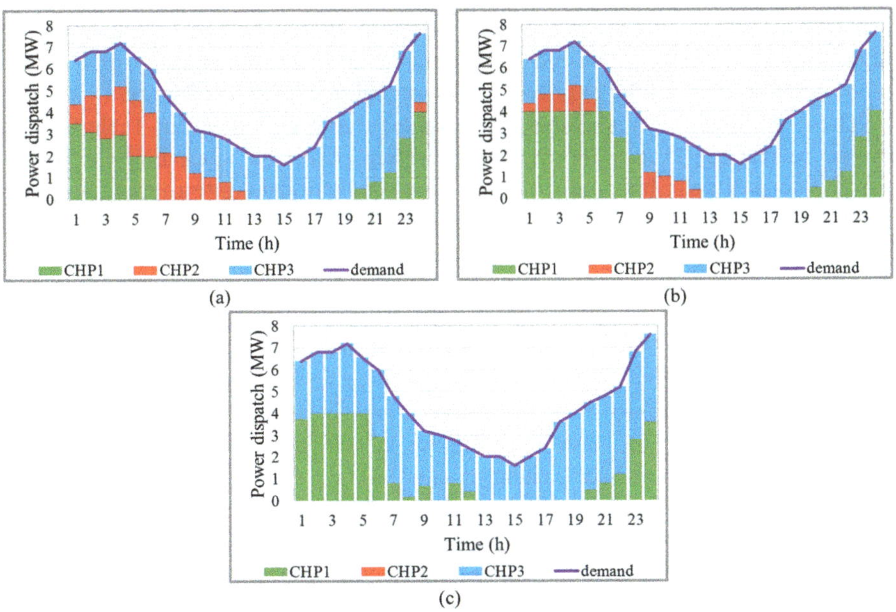

Fig. 9.7 The optimal thermal schedule of all CHPs: (**a**) Case 1. (**b**) Case 2. (**c**) Case 3

Table 9.1 The operational costs in different cases

	Cost ($)		
	Case 1	Case 2	Case 3
CHP1	2804.87	2997.59	4382.54
CHP2	4543.32	4966.89	5184
CHP3	1484.49	1480	1740.98
NGU1	1478	25	25
NGU2	4360.81	4680.78	4657.57
PV	164.83	164.83	110.59
Deg	0	1773.81	2974.43
IL	10987.58	0	0
Market	6339.79	6675.23	8146.55
Total	32019.48	22619.93	27124.91

anomaly is that with increasing robustness the operation scheduling has become far more risk-averse, thereby self-scheduling is conducted considering the possibility of higher market power and lower solar energy.

Moreover, the optimal thermal demand scheduling is illustrated in Fig. 9.7. As it is shown, in Case 1, all of CHPs are activated during hours 1–11, where the EV demand is almost zero and the thermal capacity of all units is free to be used. However, in Case 2 this demand is shifted to these hours, thereby thermal energy capacity of the CHP is reduced, as CHP2 is occupied with generating more electricity. Further, in Case 3 the solar power is diminished and market price is higher as a measure of robustness. Therefore, the electrical capacity of CHPs is more important and CHP2 is fully devoted to generate electrical energy.

These findings also resonate on cost values that are included in Table 9.1. As can be seen, Case 1 hosts the highest operation cost value since EVFs impose a massive load at peak hours and enforce high-cost units and interruptible loads. In general the cost in Case. 1 is 41% higher than that of Case 2 and 18% higher than Case 3 even though Case 3 is robust. It can be observed that market cost in Case 2 is 4% higher regarding Case 1. The reason is that higher amount of energy is imported during the hours that market energy is more affordable. Interestingly, the market price of Case 3 is 4.1% higher than that of Case 2. The reason is that this case is formulated robustly and considering lower solar power and higher market price spectrum. Therefore, more expensive energy is imported. In terms of battery degradation of the EVFs, Case 3 can impose 11% higher battery erosion since the market price is higher and solar energy is more scarce, the battery is operated under tremendously higher amount of pressure, which results in more robust and risk averse schedule. Furthermore, the production units are under more pressure in Case 2, to compensate the lower solar generation and higher market price that comes with a robust schedule.

In Table 9.2, a sensitivity analysis is conducted on the robustness of the program to show its influence on the operational cost of different equipment. Accordingly, increasing robustness can increase the operational costs exponentially. Overall, in the most robust case, the cost can be as much as 17% more. Nevertheless, it is more

Table 9.2 Cost sensitivity analysis on the robustness of the program

	$\Gamma = 0$	$\Gamma = 5$	$\Gamma = 10$	$\Gamma = 15$	$\Gamma = 20$	$\Gamma = 24$
CHP1	2997.59	3343.85	3543.86	3871.15	4056.93	4382.54
CHP2	4966.89	4988.94	5058.94	5092.85	5165.13	5184
CHP3	1480	1477.29	1484.26	1497.49	1553.47	1740.98
NGU1	25	25	25	25	25	25
NGU2	4680.79	4609.76	4708.71	4692.29	4701.55	4657.57
PV	164.83	153.98	143.13	132.28	121.44	110.59
deg	1773.81	1961.554	2131.3	2395.27	2683.06	2974.43
Market	6675.23	6979.16	7285.90	7585.62	7906.18	8146.55
Total	22619.93	23404.81	24255.88	25176.23	26106.5	27124.91

Table 9.3 Sensitivity of the problem for increasing the thermal demand of the customers

Thermal Demand (MW)	8	8.8	9.6	10.4	11.2	12
$\sum_s \sum_t \pi_s P_{h,s,t}^{CHP}, h = 1$	87.65082	87.09674	83.49312	81.18181	78.85209	76.5661
$\sum_s \sum_t \pi_s P_{h,s,t}^{CHP}, h = 2$	115.2	114.76	113.248	110.4352	107.3056	104.056
$\sum_s \sum_t \pi_s P_{h,s,t}^{CHP}, h = 3$	43.52464	48.21833	49.03377	50.0961	50.72	51.32921
$\sum_s \sum_t \pi_s P_{k,t}^{DG}, k = 1$	0.4	0.4	0.4	0.4	0.4	0.4
$\sum_s \sum_t \pi_s P_{k,t}^{DG}, k = 2$	50.62584	51.66235	53.02993	53.0999	53.07166	64.32644
$\sum_s \sum_v \sum_t \pi_s P_{s,v,t}^{PV}$	73.72855	73.72855	73.72855	73.72855	73.72855	73.72855
$\sum_s \sum_l \sum_t \pi_s P_{l,t}^{IL}$	0	0.460874	4.661017	10.7814	17.46999	20.01934
$\sum_s \sum_t \pi_s P_t^{m}$	370.7732	365.6395	364.0965	361.5541	359.3206	348.8938
$\sum_s \sum_{i,j} \sum_t \pi_s P_{s,ij,t}^{Loss}$	18.91588	18.94604	18.87957	18.70288	18.57192	17.16797
$\sum_s \sum_h \sum_t \pi_s Q_{h,s,t}^{CHP}$	106.48	117.128	127.776	138.424	149.072	159.72

reliable. For instance, the generation units are forced to have higher throughput to satisfy demand instead of the SDGs. In addition, Table 9.3 illustrates the sensitivity of the problem for increasing the thermal demand of the customers. As can be seen, increasing the thermal demand reduces the electrical production of the CHP units since the electrical and thermal production of a CHP unit is an interrelated and interconnected process. Therefore higher energy is imported from the market since CHPs cannot produce electrical energy, which is an attempt to increase thermal production. Moreover, more electrical load is interrupted since the CHPs cannot satisfy the electrical demand.

9.5 Conclusion

This chapter investigated a hybrid robust-stochastic scheduling framework for smart electric vehicle parking lots in distribution systems integrated with combined heat and power units and solar distributed generations. The electric vehicle fleets were congregated into fleets via the k-means clustering according to their uncertain parameters. The stochastic programming approach was deployed to deal with the uncertain parameters of the EVs, while the robust optimization was used to handle solar radiation and market price. The EVs were scattered on various nodes of the active distribution system (IEEE 33-bus). Furthermore, various charging strategies such as coordinated/uncoordinated charging were analyzed. Eventually, sensitivity analysis was implemented on the robustness parameter and thermal demand of the CHPs. Overall, the following outcomes were observed:

1. The smart EV charging can shift the demand, thereby improving the system operation, reducing costs and interruptible loads.
2. Uncoordinated charging is unlikely to be used in the future system as it imposes a lot of pressure on the power system.
3. The RO-SP model can offer a great deal of potential to the power system by controlling the involved risk.
4. Increasing the thermal demand of the CHPs results in severe implications as higher amount of energy needs to be imported from the market and more loads should be interrupted.

References

1. Aliasghari, P., Mohammadi-Ivatloo, B., Alipour, M., Abapour, M., & Zare, K. (2018). Optimal scheduling of plug-in electric vehicles and renewable micro-grid in energy and reserve markets considering demand response program. *Journal of Cleaner Production, 186*, 293–303. https://doi.org/10.1016/j.jclepro.2018.03.058
2. Eseye, A. T., Lehtonen, M., Tukia, T., Uimonen, S., & Millar, R. J. (2019). Optimal energy trading for renewable energy integrated building microgrids containing electric vehicles and energy storage batteries. *IEEE Access, 7*, 106092–106101. https://doi.org/10.1109/ACCESS.2019.2932461
3. Liu, Z., Chen, Y., Zhuo, R., & Jia, H. (2018). Energy storage capacity optimization for autonomy microgrid considering CHP and EV scheduling. *Applied Energy, 210*, 1113–1125. https://doi.org/10.1016/j.apenergy.2017.07.002
4. Razipour, R., Moghaddas-Tafreshi, S. M., & Farhadi, P. (2019). Optimal management of electric vehicles in an intelligent parking lot in the presence of hydrogen storage system. *Journal of Energy Storage*. https://doi.org/10.1016/j.est.2019.02.001
5. Jannati, J., & Nazarpour, D. (2017). Optimal energy management of the smart parking lot under demand response program in the presence of the electrolyser and fuel cell as hydrogen storage system. *Energy Conversion and Management, 138*, 659–669. https://doi.org/10.1016/j.enconman.2017.02.030
6. Marzoghi, A. F., Bahramara, S., Adabi, F., & Nojavan, S. (2019). Optimal scheduling of intelligent parking lot using interval optimization method in the presence of the electrolyser

and fuel cell as hydrogen storage system. *International Journal of Hydrogen Energy, 44*(45), 24997–25009. https://doi.org/10.1016/j.ijhydene.2019.07.226
7. Liu, J., Chen, C., Liu, Z., Jermsittiparsert, K., & Ghadimi, N. (2020). An IGDT-based risk-involved optimal bidding strategy for hydrogen storage-based intelligent parking lot of electric vehicles. *Journal of Energy Storage, 27*, 101057. https://doi.org/10.1016/j.est.2019.101057
8. Sengor, I., Erdinc, O., Yener, B., Tascikaraoglu, A., & Catalao, J. P. S. (2019). Optimal energy management of EV parking lots under peak load reduction based DR programs considering uncertainty. *IEEE Transaction on Sustainable Energy, 10*(3), 1034–1043. https://doi.org/10.1109/TSTE.2018.2859186
9. Zeynali, S., Nasiri, N., Marzband, M., & Ravadanegh, S. N. (2021). A hybrid robust-stochastic framework for strategic scheduling of integrated wind farm and plug-in hybrid electric vehicle fleets. *Applied Energy, 300*(xxxx), 117432. https://doi.org/10.1016/j.apenergy.2021.117432
10. Wei, Z., Peng, K., Chen, J., Yan, X., & Wan, Q. (2019). Stability analysis of a DC distribution system for power system integration of plug-in electric vehicles. *2019 IEEE PES Innov. Smart Grid Technol. Asia, ISGT 2019*, 2450–2455. https://doi.org/10.1109/ISGT-Asia.2019.8881468
11. Liu, X. (2020). Research on flexibility evaluation method of distribution system based on renewable energy and electric vehicles. *IEEE Access, 8*, 109249–109265. https://doi.org/10.1109/ACCESS.2020.3000685
12. Rahimi, K., & Davoudi, M. (2018). Electric vehicles for improving resilience of distribution systems. *Sustainable Cities and Society, 36*, 246–256. https://doi.org/10.1016/j.scs.2017.10.006
13. Monteiro, V., Gonçalves, H., & Afonso, J. L. (2011). Impact of electric vehicles on power quality in a Smart Grid context. *Proceeding of International Conference on Electric Power Quality and Utility EPQU*, 660–665. https://doi.org/10.1109/EPQU.2011.6128861
14. Sadeghian, O., Nazari-Heris, M., Abapour, M., Taheri, S. S., & Zare, K. (2019). Improving reliability of distribution networks using plug-in electric vehicles and demand response. *Journal of Modern Power System and Clean Energy, 7*(5), 1189–1199. https://doi.org/10.1007/s40565-019-0523-8
15. Ul-Haq, A., Cecati, C., & El-Saadany, E. (2018). Probabilistic modeling of electric vehicle charging pattern in a residential distribution network. *Electric Power Systems Research, 157*, 126–133. https://doi.org/10.1016/j.epsr.2017.12.005
16. Nour, M., Ramadan, H., Ali, A., & Farkas, C. (2018). Impacts of plug-in electric vehicles charging on low voltage distribution network. *Proceedings of 2018 International Conference on Innovative Trends Computing Engineering ITCE 2018, 2018*, 357–362. https://doi.org/10.1109/ITCE.2018.8316650
17. Pirouzi, S., Latify, M. A., & Yousefi, G. R. (2020). Conjugate active and reactive power management in a smart distribution network through electric vehicles: A mixed integer-linear programming model. *Sustainable Energy, Grids Networks, 22*, 100344. https://doi.org/10.1016/j.segan.2020.100344
18. Galiveeti, H. R., Goswami, A. K., & Dev Choudhury, N. B. (2018). Impact of plug-in electric vehicles and distributed generation on reliability of distribution systems. *International Journal of Engineering Science and Technology, 21*(1), 50–59. https://doi.org/10.1016/j.jestch.2018.01.005
19. Cheng, S., & Li, Z. (2019). Multi-objective network reconfiguration considering v2g of electric vehicles in distribution system with renewable energy. *Energy Procedia, 158*, 278–283. https://doi.org/10.1016/j.egypro.2019.01.089
20. Nasiri, N., Zeynali, S., Ravadanegh, S. N., & Marzband, M. (2021). A hybrid robust-stochastic approach for strategic scheduling of a multi-energy system as a price-maker player in day-ahead wholesale market. *Energy, 235*. https://doi.org/10.1016/j.energy.2021.121398
21. Zeynali, S., Rostami, N., Ahmadian, A., & Elkamel, A. (2021). Robust multi-objective thermal and electrical energy hub management integrating hybrid battery-compressed air energy storage systems and plug-in-electric-vehicle-based demand response. *Journal of Energy Storage, 35*, 102265. https://doi.org/10.1016/j.est.2021.102265

22. Banzo, M., & Ramos, A. (2011). Stochastic optimization model for electric power system planning of offshore wind farms. *IEEE Transactions on Power Apparatus and Systems, 26*(3), 1338–1348. https://doi.org/10.1109/TPWRS.2010.2075944
23. Bisschop, J. (2006). *AIMMS optimization modeling*.
24. Bahramara, S., Sheikhahmadi, P., Mazza, A., Chicco, G., Shafie-Khah, M., & Catalão, J. P. S. (2020). A risk-based decision framework for the distribution company in mutual interaction with the wholesale day-ahead market and microgrids. *IEEE Transaction on Industrial Informatics, 16*(2), 764–778. https://doi.org/10.1109/TII.2019.2921790
25. Nasiri, N., et al. (2020). A bi-level market-clearing for coordinated regional-local multi-carrier systems in presence of energy storage technologies. *Sustainable Cities and Society, 63*, 102439. https://doi.org/10.1016/j.scs.2020.102439
26. Sheikhahmadi, P., Bahramara, S., Mazza, A., Chicco, G., & Catalão, J. P. S. (2021). Bi-level optimization model for the coordination between transmission and distribution systems interacting with local energy markets. *International Journal of Electrical Power & Energy Systems, 124*, 106392. https://doi.org/10.1016/j.ijepes.2020.106392
27. Zeynali, S., Rostami, N., Feyzi, M. R., & Mohammadi-ivatloo, B. (2020). Multi-objective optimal planning of wind distributed generation considering uncertainty and different penetration level of plug-in electric vehicles. *Sustainable Cities and Society, 62*, 102401. https://doi.org/10.1016/j.scs.2020.102401
28. Xiong, Y., Wang, B., Chu, C. C., & Gadh, R. (2018). *Electric vehicle driver clustering using statistical model and machine learning* (Vol. 2018). IEEE Power Energy Society General Meeting. https://doi.org/10.1109/PESGM.2018.8586132

Chapter 10
Operational Challenges of Electric Vehicle Smart Charging

Hamdi Abdi, Maryam Shahbazitabar, and Mansour Moradi

10.1 Status Quo, Challenges, and Outlook

Over the past decades, electric vehicles (EVs) have become one of the new technologies in achieving ambitious clean energy and decarbonization because conventional transportation systems are recognized as the largest source of environmental pollution due to the consumption of about 62.3% of the world's fossil fuels [1]. According to a new IEA report published on April 20, 2021, the emission of energy-related carbon dioxide in the world to date has increased to 1.5 billion tons, an increase of 60.97% compared to 1990. This is the second-largest increase in energy-related carbon dioxide emissions globally since the severe financial downturn of 2010, which was also very significant. But in 2020, due to the Covid-19 pandemic and the decline in energy use, especially in the transportation sector, this increase was reduced by 5.69% to 2019. However, in 2021, this decrease was again increased, as depicted in Fig. 10.1 [2].

A large part of the greenhouse gas (GHG) emission is generated from the vehicle's internal combustion engines. Motor vehicles account for about 16% of the global emission of man-made carbon dioxide [3, 4], reduce air quality, and threaten social health significantly. In this regard, the European Union has presented a roadmap for 2050 with 80% reduction in GHG emissions compared to 1990 [1].

H. Abdi (✉) · M. Shahbazitabar
Electrical Engineering Department, Engineering Faulty, Razi University, Kermanshah, Iran
e-mail: hamdiabdi@razi.ac.ir; m.sht@razi.ac.ir

M. Moradi
Department of Electrical Engineering, Technical and Vocational University (TVU), Kermanshah, Iran
e-mail: Mansour.Moradi.ir@gmail.com

V. Vahidinasab, B. Mohammadi-Ivatloo (eds.), *Electric Vehicle Integration via Smart Charging*, Green Energy and Technology,
https://doi.org/10.1007/978-3-031-05909-4_10

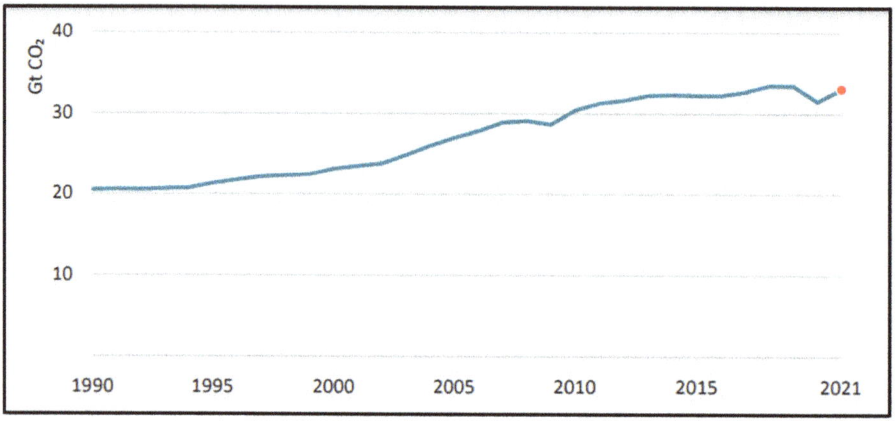

(a)

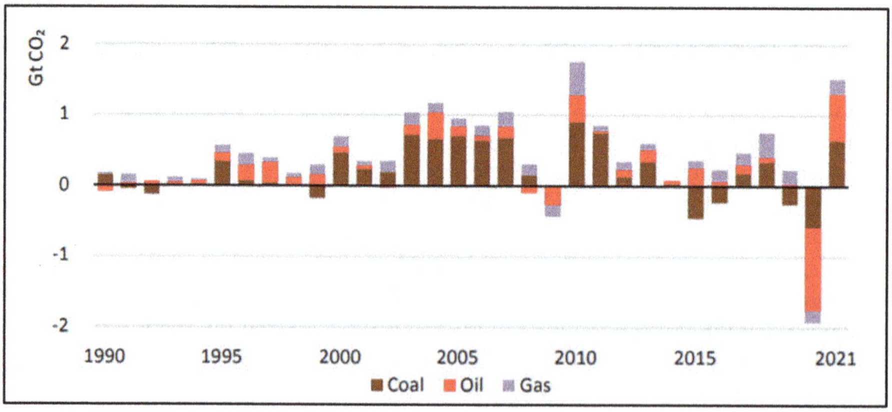

(b)

Fig. 10.1 (**a**) A global energy-related CO_2 emission, 1990–2021 [2]. (**b**) Changes in CO_2 emissions from fuel, 1990–2021 [2]

In this regard, some EU countries have imposed restrictions such as a ban on the sale of classical cars until 2040 and then a ban on their driving until 2050 [5, 6].

EVs are the most promising vehicles of the twenty-first century, which have great effects on pollution reduction and economic issues [7, 8]. Generally, not only EVs can reduce the use of fossil fuels, greenhouse gas emissions, and environmental pollution, but also can contribute to bidirectional energy exchange by attracting/injecting the electrical energy from/to the grid at different times (G2V-V2G). However, widespread penetration of EVs is associated with adverse effects on local distribution networks [9]. Most of these effects are due to uncoordinated charging, which leads to increased power demand at peak hours. In addition, the penetration of large-scale EVs has potentially led to wide-range changes in aggregate demand that cannot be easily controlled by the existing distribution network infrastructure [10]. Therefore, the smart charge management system (SCMS) will control the charging process in public and private charging stations by shifting loads to more desirable times.

Smart charging allows customers and network operators to schedule charging profiles of plug-in hybrid electric vehicles (PEVs) to get technical and economic benefits, considering a specific demand-side management (DSM) of PEVs. It means, smart charging seeks active control of loads and can be programmed with optimization methods to achieve profit for both grid and consumer such as reducing generation costs and GHG emissions, frequency regulation and voltage control, valley filling and peak shaving, spinning reserve and emergency demand response, integration of renewable energy source, etc. Therefore, considering the importance of this issue, this chapter will review the definitions, types of EV technologies, different types of charging, advantages and challenges, and relevant standards in this regard.

10.2 Definition

An optimal intelligent strategy for bidirectional power exchange controls should be adopted for charge and discharge of the EVs connected to the network. Therefore, smart charging (SC) means that a user can intelligently manage the charging method by connecting the EV to the main grid. Thus, smart charging allows the operator to control the energy required for each EV, and to set specific limits on power consumption (whether one has a charger in one's home or owns a business with multiple charging stations). Furthermore, SCMS should optimize the charging/discharging of EVs at public charging stations and residential charging stations. In addition, this system either is able to transfer the network loads to more favorable times, and also participate in services such as voltage and frequency support and integration of renewable energy sources. Therefore, to provide multiple services, SCMS must properly control the EVs connected to the electric vehicle supply equipment (EVSEs). On the other hand, it is necessary to monitor and control the data connection between EVs and EVSEs, which is done in real-time with SCMS and the network. For this purpose, SCMS can be susceptible to cyber-physical attacks. Therefore, the cyber-physical security of SCMS must be properly monitored to ensure that the system is secure against attacks to reduce the number of attacks on charging points and customers [11].

Therefore, it is predicted that the growth of the penetration of these vehicles will continue due to the following reasons [3–6]:

- Promoting clean-fuel vehicles and striving for carbon offsets: Many countries have specific incentive policies for clean fuel vehicles. For example, the Netherlands and France will ban fossil fuel vehicles from 2025 and 2040, respectively [3].

- Reducing the anxiety of EV drivers in the field of driving without recharging: Anxiety in the traversable range is known as the main obstacle in accepting EVs [4].
- EV wireless charging experience: One of the crucial features of EVSC is the deployment of remote-control systems through smartphone applications. This technology not only makes EV charging faster but also user-friendly.

This type of charging allows the charging station management to remotely monitor, manage, and optimize energy efficiency and, if necessary, impose restrictions on the use of charging stations under its control. In this case, EVs connected to the grid with intelligent charging cause a proper coexistence with the power grid [4].

10.3 Electric Vehicle Technology

Various EVs are connected to local distribution networks in terms of manufacturing technology and classified into several different categories [11–13]:

- *Battery electric vehicle (BEV):* These vehicles use batteries to supply the electrical energy for the car's electric motor, as well as other equipment.
- *Hybrid electric vehicles (HEVs):* Unlike EVs, these vehicles have more than one power source. These types of cars have both combustion engines and electric motors with sufficient batteries to store energy from the combustion engine and car brakes. The internal combustion engine consists of a fuel tank that transmits the combustion output to the wheels via a power transmission system. In contrast, the electric motor consists of batteries that transmit energy stored in the batteries to the wheels via a power transmission system when needed.
- *Plug-in hybrid electric vehicles (PHEVs):* These types of vehicles, which are designed to eliminate the disadvantages of hybrid electric vehicles, can also be charged by the grid. These cars have a complete combustion engine system. In these cars, the electric motor is used in driving in the city where the distances are shorter, and the number of starts and stops of the car is more, and in intercity driving where the distance will be longer, the combustion engine is used.

BEVs and PHEVs are commonly used to classify plug-in electric vehicles (PEVs), as both types must be connected to the main network to charge their batteries.

10.4 Electric Vehicles Charging

Depending on the different technologies used in EVs, electric vehicles can be charged using transmission, induction, or battery charging stations. However, the investments are different for each of the battery charge modes.

Table 10.1 Charging level of electric vehicles and charging standards [14, 15]

Charging level	Voltage level	Range	Power (kW)	Application
Level I 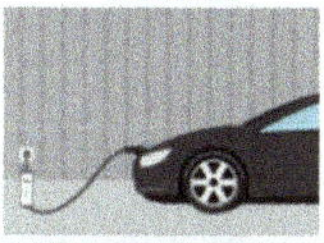	120 VAC	2–5 miles of range per hour	1.4–1.9	Homes Multi-unit residential
Level II 	240 VAC	10–30 miles of range per hour	2.5–19.2	Homes Multi-unit residential Workplace Fleet Public
DC 	480 VDC	150–350+ miles of range per hour	50–150	Multi-unit residential Fleet Public

10.4.1 Charging Standards for Electric Vehicles

EVs are usually charged by a dedicated charging circuit. Charging is usually done by 120 V (level 1) or 240 V (level 2). Level one is used for 16 Amps and above and level 2 is used for 32 Amps and above. In some countries, charging is done by 230 V single-phase or 400 V three-phase electricity, which is done by IEC-62196 and SAE-J1772 standards [14, 15]. For fast charging, they usually install locations near highways or crowded parking lots according to IEC Standard 61851 [16]. Table 10.1 presents the charge level and charging standards of electric vehicles.

Efficiency, capacity, and charging speed are the other challenges in EVs technology. Although new batteries can give all their stored power in less time, high-capacity and fast-charging batteries require a lot of power, which must be provided by the distribution system or power grid.

10.4.2 Charging Speed and Duration

There are many different types of EVs looking to the future in different abilities, functionalities, and capacities growth. To charge these cars, two modes are considered, fast charging and slow charging [17]. The characteristics of these two charging modes are based on the charge hours and the power that each vehicle receives from the network during charging, as shown in Table 10.2. With ultra-fast charging technology, car charging time is reduced from hours to minutes. Using this type of charge makes the charging time of batteries 10 times faster than ordinary EV batteries today [18].

Table 10.2 Battery charging time [17, 18]

Charging type	Charging time (h)	Vehicle consumption during charging (kW)
Slow	6	3.7
Fast	2	11
Ultra-fast	0.1	60–500

On the other hand, the capacity of installed batteries in EVs is not all the same. The capacity of each battery is measured in kilowatt-hours (kWh). The higher the kilowatt-hour of a battery, the more energy can be stored, and therefore the longer the car can travel on a single full charge. But in general, ideally, this amount of charge can be equal to the power of the charging station. But it is almost restricted by a set of factors, including charging station power, maximum charging power of the device, maximum current tolerable by the charger cable, and network energy availability in real [19].

To maintain system reliability and better control, a limited number of vehicles can be charged and discharged at a specific time interval. The number of times for charge and discharge depends on the type of battery and its lifetime, and it is usually taken once a day for simplicity. In batteries, the state of charge (SOC) is related to the type of battery and indicates the amplitude of the current flowing in/out of the battery at the time of charging/discharging.

Four different modes are defined for charging EVs as follows [20]:

- Uncontrolled charging Mode-UCM: The owners of these cars charge their appliances in an uncontrolled manner as soon as they reach home, which is usually in the early hours of the night at peak load conditions. They connect their car to the main network and the charging process begins and stops with a full charge.
- Delayed charging Mode-DCM: This type of charging is also done at home, but to use the low cost in the load response program, it will be done from 10 p.m. onwards.
- Off-peak charging Mode-OCM: This mode is similar to the previous one, except that the start and the end of the charge are directly, or indirectly controlled by the central control system.
- Continues charging Mode-CCM: In this type of charging, cars are charged continuously and wherever they are stopped, whether at home or public charging stations.

Therefore, all vehicles registered in the network must participate in the network around the clock.

10.4.3 Electric Vehicle Smart Charging (EVSC)

In general, vehicles are ideally in 95% of the time in standby mode, and users need only 10% of their daily working hours to charge [13, 21]. This allows flexible services to be provided to customers at least 85% of the time. But by changing the pattern of mobility, an increase in the duration of use of vehicles through new and common smart mobility services can be seen. Therefore, with proper and optimal planning of EVSC, in 85% of the remaining time to provide flexible services, the charging process can be well controlled [13, 22–25].

EVSC refers to a system in which an EV and a charging device establish a data connection via cloud technology, allowing the charging station owner to remotely monitor and manage energy efficiency and limitations to use their devices. If significant numbers of EVs start charging at the same time, the system may encounter disruptions. EVSC will act as a key element when connecting EVs to the grid [26, 27].

EVSC is offered in the following three models as shown in Fig. 10.2 [28]:

- One-way and controlled charging is called V1G. In this model, according to the charging infrastructure, the charging amount is adjusted based on network requirements.
- Charging the EVs to the network, which is called V2G. In this model, at peak load when the network is experiencing power shortages, the batteries of EVs are used as a source to supply some portions of this power [29].
- Charging a car at home or building is called V2H-B. In this model, the energy stored in the car battery is used based on the immediate energy needs of the house or building and is recharged by reducing the amount of demand during off-peak hours.

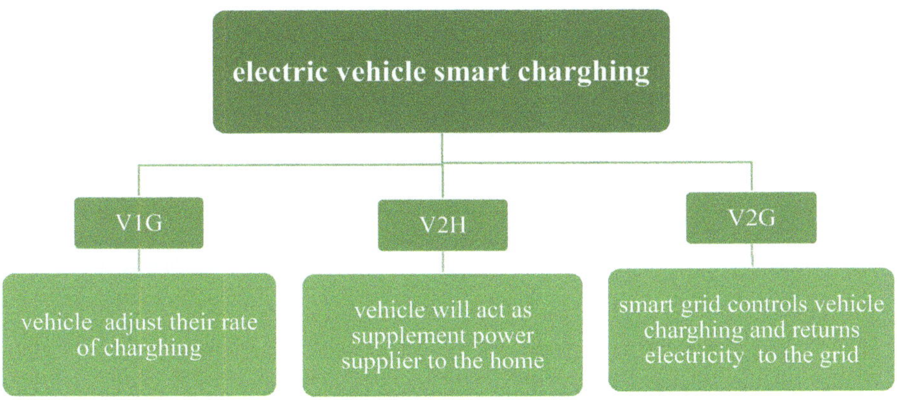

Fig. 10.2 Electric vehicle smart charging models

The charging infrastructure in the V2G and V2B-H models continuously manage the demand of the network or building to ensure that the vehicles and the network benefit from the existing interaction and none of them malfunction.

10.5 Control of EVSC: Centralized and Decentralized Control Approaches

Generally, based on existing studies, EVs charge control schemes are classified into two modes of centralized and decentralized schemes, each of which depends on the type of information available and the relationship between EV and the network [13, 30].

- *Centralized charge control:* In this strategy, a central unit collects network, load, and generation information to decide whether to charge or discharge the EV fleet. Therefore, the centralized control approach requires a significant communication infrastructure.
- *Decentralized charge control:* A distributed charging strategy in which the owner of each EV decides when to charge the battery, assuming there is no central controller.

Centralized charge control is impossible due to poor scalability and the need to know user behavior, while decentralized charge control strategies are scalable but suffer from EV behavior uncertainties. Decentralized control, on the other hand, will require accurate and predictive consumer reaction, which is not possible in an environment with dynamic EVs [30–32]. In general, the advantages and disadvantages of each charge control scheme are presented in Table 10.3 [30, 32].

Table 10.3 Advantages and disadvantages of each EVs charge control scheme [30, 32]

Control type	Advantages	Disadvantages
Centralized	Well known architecture, better ancillary services provision, and better utilization of network capacity	Expensive and complex communication infrastructure, need the central controller and a backup system, increasing the complexity with the number of PEVs, need a large amount of data processing, and possible privacy violations
Decentralized	Scalable, fewer communication infrastructures, improve fault tolerance, higher consumer acceptance, and charge control remain in the user	Uncertain results, limited ancillary services provision, need the forecasting or predicting the reaction of consumers, probable simultaneous reactions or avalanche effects

10.6 Benefits of EVSC

Figure 10.3 shows some significant benefits of ECSC to the power system to which they are related in any way.

From the point of view of equipment, smart energy management can optimize the power of their infrastructure by efficiently distributing the available power between cars and other consumers. From a long-term investment perspective, it brings operational benefits to various EVSC-related occupations. Depending on the type of electricity storage infrastructure, they can supply the required power to the grid, and return their excess power to the grid in times of peak demand. In addition, smart charging can increase the building's energy capacity while vehicles can act as an electrical storage system. EV owners can also increase the benefits by integrating home renewable energy systems and increase total costs by storing extra energy in their vehicles during the day, and then draining the energy when needed or at night for use in their homes [28, 33]. Overall, this system can be considered as the biggest incentive for EV drivers to get cheaper, environmentally friendly, and safer charging.

The following five general advantages can also be considered for this type of charge [33, 34]:

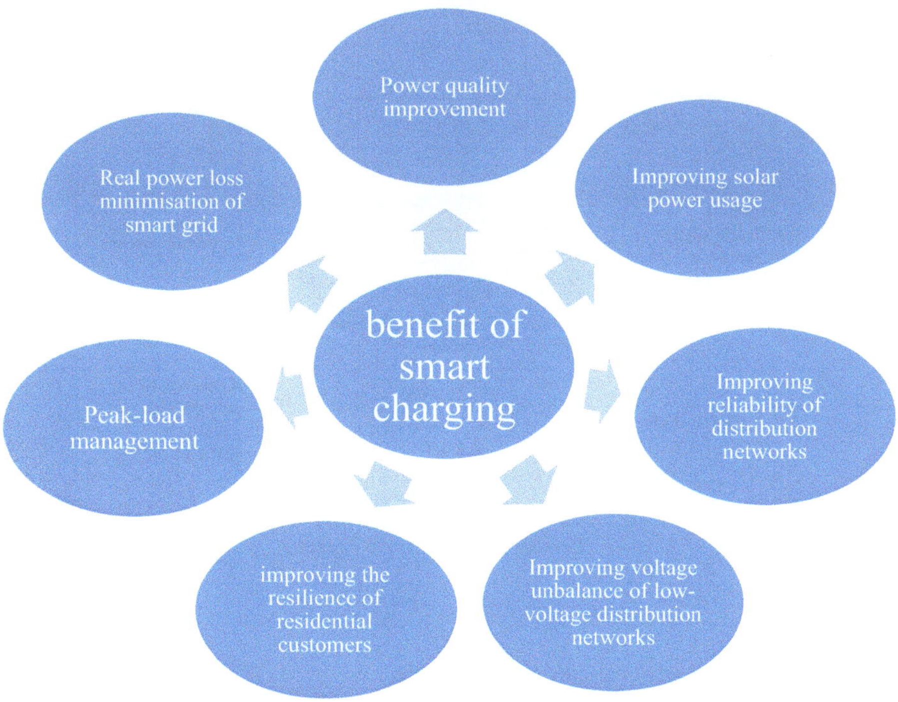

Fig. 10.3 Some significant benefits of ECSC

- Charge faster,
- Safer charging
- Monitoring electricity consumption
- Charging time optimization
- Find charging stations

10.7 Main Challenges of Using EVSCs

Some of the electric vehicle challenges are shown in Fig. 10.4. One of the main challenges for EVSCs is their potential consequences in terms of their maximum instantaneous charging capacity. This will happen when all the EVs start charging simultaneously as soon as they reach their destination. This increases the risk in the network and a lot of investment must be made in the network infrastructure to make charging smarter.

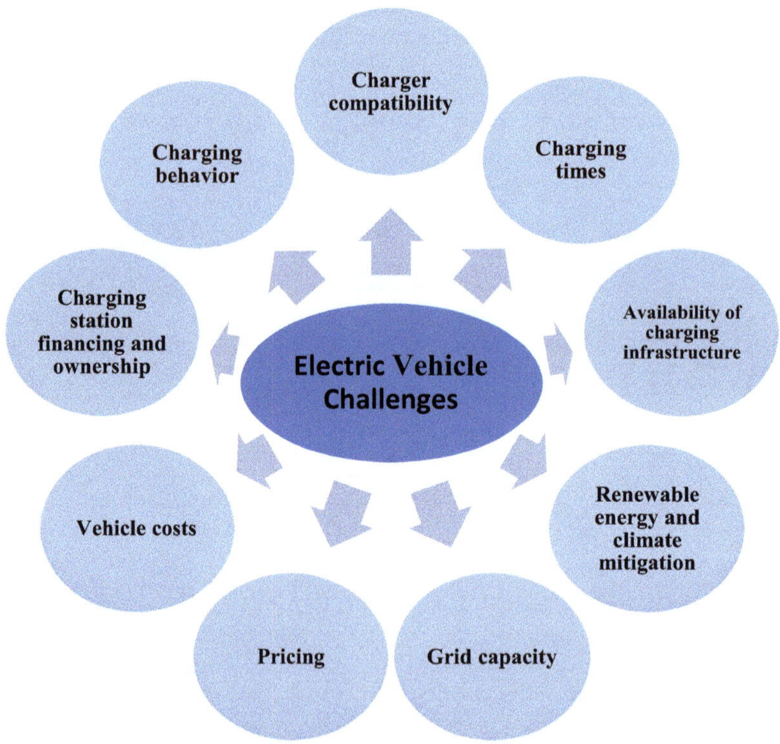

Fig. 10.4 Main challenges of electric vehicle charging

But a significant challenge in this area is related to smart charging software. Because the energy infrastructure is currently available with a network consisting of cables, transformers, and other electrical equipment, an additional layer of software network infrastructure will be needed to determine how much energy, at what time, and where it should be available. This is considered as the main challenge because many variables play a role in energy allocation that will be very difficult to coordinate [35].

On the consumer side, there are many loads, such as home appliances, which cause uncertainty in the network. Also, when using renewable energy sources, there are many uncertainties regarding the availability of their output power. In addition, there are various variables that affect the power grid and make it difficult to plan for the supply and allocation of energy for such an EVSC [35].

10.7.1 Connectivity and Infrastructure in EVSC

Worldwide concerns on environmental pollution, global warming, social welfare, and lifestyle quality are some important reasons that motivate cities to become smarter and more intelligent. One of these attempts refers to smart transportation and smart charging, providing regional development and technology utilization, especially smart-based technologies, by using information and communication technology, and services [36].

This has dramatically accelerated research into EVSC infrastructure management at breakneck speed. Because ensuring the availability of recharge ability at the network level, it is necessary to create an extensive public charging infrastructure. This issue is considered as an interdisciplinary field and the issue has made research more complex and evaluates the field of collective research management. Also, EVs will not be a viable option for drivers unless a large-scale charging infrastructure is available. On the other hand, as long as there are not enough electric vehicles, it is doubtful that charging service providers will invest heavily in infrastructure [37].

Due to the existing problems in this field in the last few years, various researchers have studied it. Many of these studies have examined the characteristics of EV refueling in their optimization models for determining and sizing EV charging infrastructure. A summary of the research, including the purpose of the study, optimization model, optimization solution method, and evaluation method for each of them, is presented in the literature [38].

10.7.2 The Minimum Requirements for EVSC

The minimum requirements that must be met in each of the EVSC are as follows [39]:

1. For the smart charger to be able to support a wide range of different smart applications, the minimum components required for better performance at smart charging points can be summarized as:
   - The watch.
   - Information storage system.
   - Electricity meter to meet the needs of the program.
   - Network interface to be able to support remote.
   - Load control calendar. It saves the daily load for the charging point and can be changed continuously. This section, based on the regulatory calendar, allows the charger to continue working according to the previous schedule in case of disconnection with the operator.
   - A proportionate controller that allows the charging point to control the current according to the load control calendar.
2. The smart charger must be able to operate within the limits imposed by the local distribution system operators.
3. The performance of the smart charger during and after the error must be determined, planned, and agreed upon in advance.
4. When charging/discharging control, the smart charger must use a random compensator to change power events.
5. Having the ability of local control.
6. Be able to reuse power.
7. In smart charging, cyber risk and security assessment should be applied to be able to fulfill its obligations fully.
8. Any organization that will be charged for smart cyber security monitoring should conduct a review by international standards and will not identify a preferred option that can be supported. This eliminates the need for the charge point operators (CPO) to justify the choice of standard and to provide an incentive scheme for the subscriber with an adaptive approach.
9. An expert group should create organizational requirements related to cyber security in proportion to controlled-load for intelligent charging, including ISO 15118.
10. The actors of the smart charging process should pay attention to data privacy issues and, if necessary, impose fines for non-compliance with data protection laws.

10.8 Conclusion

Electric vehicles are essential contributors to the green world and participate in pollution and greenhouse gas emission reduction. The electric grid is faced challenges in frequency and voltage drop because of the bidirectional relation between the grid and PHEVs. So, it is vital to optimize the charging and discharging schedule considering all the environmental, physical, and social constraints with EVSC to overcome uncertainty. Centralized or decentralized control, public or residential charging station, delayed or fast charging, owner behavior, battery specification, and demand-side consideration are some of the main and basic components in EVSC designing which must be considered economically.

References

1. Sadati, S. M. B., et al. (2019). Operational scheduling of a smart distribution system considering electric vehicles parking lot: A bi-level approach. *International Journal of Electrical Power & Energy Systems, 105*, 159–178.
2. International Energy Agency (IEA). *Global energy review 2021. CO2 emissions*. https://www.iea.org/reports/global-energy-review-2021/co2-emissions
3. Climate change and CO2. International Organization of Motor Vehicle Manufacturers, Paris, France, May 2008.
4. Kong, P.-Y., & Karagiannidis, G. K. (2016). Charging schemes for plug-in hybrid electric vehicles in smart grid: A survey. *IEEE Access, 4*, 6846–6875.
5. European Commission. (2011). *Communication from the Commission to the European Parliament, the Council, the European Economic and Social Committee and the Committee of the Regions. Energy Roadmap 2050*. European Commission.
6. Saldaña, G., et al. (2019). Electric vehicle into the grid: Charging methodologies aimed at providing ancillary services considering battery degradation. *Energies, 12*(12), 2443.
7. Mohamed, M., et al. (2016). Identifying and characterizing potential electric vehicle adopters in Canada: A two-stage modelling approach. *Transport Policy, 52*, 100–112.
8. Xiao, H., et al. (2014). A survey of influence of electrics vehicle charging on power grid. In *2014 9th IEEE conference on Industrial Electronics and Applications*. IEEE.
9. Dharmakeerthi, C. H., Mithulananthan, N., & Saha, T. K. (2014). Impact of electric vehicle fast charging on power system voltage stability. *International Journal of Electrical Power & Energy Systems, 57*, 241–249.
10. Qian, K., Zhou, C., & Yuan, Y. (2015). Impacts of high penetration level of fully electric vehicles charging loads on the thermal ageing of power transformers. *International Journal of Electrical Power & Energy Systems, 65*, 102–112.
11. Bhusal, N., Gautam, M., & Benidris, M. (2021). Cybersecurity of electric vehicle smart charging management systems. In *2020 52nd North American Power Symposium (NAPS)*. IEEE.
12. Hildermeier, J., et al. (2019). Smart EV charging: A global review of promising practices. *World Electric Vehicle Journal, 10*(4), 80.
13. Saldanha, J. J. A., et al. (2016). Control strategies for smart charging and discharging of plug-in electric vehicles. *Smart Cities Technologies, 1*, 121–141.
14. Plugs, socket-outlets, vehicle connectors and vehicle inlets—conductive charging of electric vehicles, IEC Standard 62196, 2014.
15. SAE electric vehicle and plug in hybrid electric vehicle conductive charge coupler, SAE Standard J1772, 2013.
16. Electric vehicle conductive charging system, IEC Standard 61851, 2011.
17. Yang, Y., Tan, Z., & Ren, Y. (2020). Research on factors that influence the fast charging behavior of private battery electric vehicles. *Sustainability, 12*(8), 3439.
18. https://www.batterypoweronline.com/markets/charging/proterra-introduces-high-power-interoperable-ev-charging-technology/
19. https://www.dazetechnology.com/how-long-does-it-take-to-charge-an-electric-vehicle/

20. Parks, K., Denholm, P., & Markel, T. (2007). *Costs and emissions associated with plug-in hybrid electric vehicle charging in the Xcel Energy Colorado service territory. No. NREL/TP-640-41410*. National Renewable Energy Lab. (NREL), Golden, CO (United States).
21. Langton, A., & Crisostomo, N. (2013). *Vehicle-grid integration: A vision for zero emission transportation interconnected throughout California's electricity system*. California Public Utilities Commission: San Francisco, CA, USA. Available online: http://docs.cpuc.ca.gov/PublishedDocs/Published/G000/M080/K775/80775679.pdf. Accessed on 20 Sep 2019.
22. Canzler, W., & Knie, A. (2016). Mobility in the age of digital modernity: Why the private car is losing its significance, intermodal transport is winning and why digitalization is the key. *Applied Mobilities., 1*, 56–67.
23. Hochfeld, C., Jung, A., Klein-Hitpaß, A., Maier, U., Meyer, K., & Vorholz, F. (2017). *Mit der Verkehrswende die Mobilität von Morgen Sichern. 12 Thesen zur Verkehrswende*. Agora Verkehrswende: Berlin, Germany. Available online: https://www.agora-verkehrswende.de/fileadmin/Projekte/2017/12_Thesen/Agora-Verkehrswende-12-Thesen-Kurzfassung_WEB.pdf. Accessed on 20 Sep 2019.
24. Slowik, P., & Kamakate, F. (2017). *New mobility: Today's technology and policy landscape*. The International Council for Clean Transportation: Washington, DC, USA. Available online: https://theicct.org/sites/default/files/publications/New-mobility-landscape_ICCT-white-paper_27072017_vF.pdf. Accessed on 20 Sep 2019.
25. Beckmann, K. J., & Klein-Hitpaß, A. (Eds.). (2013). *Nicht Weniger Unterwegs, Sondern Intelligenter? Neue Mobilitätskonzeptei*. German Institute of Urban Affairs (Difu): Berlin, Germany. Available online: https://difu.de/publikationen/2013/nicht-weniger-unterwegs-sondern-intelligenter.html. Accessed on 20 Sep 2019.
26. European Union. *Energy Union: Commission takes action to reinforce EU's global leadership in clean vehicles*. Available online: https://ec.europa.eu/transport/modes/road/news/2017-11-08-driving-cleanmobility_en. Accessed on 20 Sep 2019.
27. Buck, M., Graf, A., & Ten Graichen, P. (2019). Priorities for the next European Commission to meet the EU's 2030 targets and accelerate towards 2050. In *European energy transition 2030: The big picture*. Agora Energiewende: Berlin, Germany. Available online: https://www.agora-energiewende.de/fileadmin2/Projekte/2019/EU_Big_Picture/153_EU-Big-Pic_WEB.pdf. Accessed on 20 Sep 2019.
28. https://driivz.com/blog/ev-smart-charging-benefits/
29. Shahbazitabar, M., & Abdi, H. (2018). A novel priority-based stochastic unit commitment considering renewable energy sources and parking lot cooperation. *Energy, 161*, 308–324.
30. Abdullah, H. M., Gastli, A., & Ben-Brahim, L. (2021). Reinforcement learning based EV charging management systems–a review. *IEEE Access, 9*, 41506–41531.
31. García-Villalobos, J., et al. (2014). Plug-in electric vehicles in electric distribution networks: A review of smart charging approaches. *Renewable and Sustainable Energy Reviews, 38*, 717–731.
32. Veneri, O. (Ed.). (2017). *Technologies and applications for smart charging of electric and plug-in hybrid vehicles*. Springer International Publishing.
33. IRENA. (2019). *Innovation landscape brief: Electric-vehicle smart charging*. International Renewable Energy Agency.
34. https://www.virta.global/blog/benefits-of-smart-charging
35. European Union's Horizon 2020 research and innovation programme under grant agreement No 769016.
36. Abdi, H., & Shahbazitabar, M. (2020). Smart city: A review on concepts, definitions, standards, experiments, and challenges. *Journal of Energy Management and Technology, 4*(3), 1–6.
37. Ustun, T. S., et al. (2021). IEC-61850-based communication for integrated EV management in power systems with renewable penetration. *Energies, 14*(9), 2493.
38. Shen, Z.-J. M., et al. (2019). Optimization models for electric vehicle service operations: A literature review. *Transportation Research Part B: Methodological, 128*, 462–477.
39. Work package three: Smart charging technical requirements. 14 October 2020.

Index

V. Vahidinasab, B. Mohammadi-Ivatloo (eds.), *Electric Vehicle Integration via Smart Charging*, Green Energy and Technology,
https://doi.org/10.1007/978-3-031-05909-4

Ingram Content Group UK Ltd.
Milton Keynes UK
UKHW021951020723
424413UK00001B/7

9 783031 059087